CANADA'S DEADLY SECRET

Other works by Jim Harding

"Pharmaceutical Control" (three-part series, CBC *Ideas*, 1983)

Social Policy and Social Justice (Wilfred Laurier University Press, 1995)

After Iraq: War, Imperialism and Democracy (Fernwood Publishing, 2004)

Student Radicalism and National Liberation: Essays on the "New Left" Revolt in Canada 1964–74 (Crows Nest, 2006)

Between Ages: Ancestral Revelations, Personal Awakenings and Spiritual Possibilities (forthcoming)

CANADA'S DEADLY SECRET

Saskatchewan Uranium and the Global Nuclear System

Jim Harding

Fernwood Publishing • Halifax and Winnipeg

Editing: Richard Therrian and Brenda Conroy
Cover Design: John van der Woude
Dedication photo: Gary Robbins, back cover photo: Marie Buga
Printed and bound in Canada by Hignell Book Printing

Published in Canada by Fernwood Publishing
Site 2A, Box 5, 32 Oceanvista Lane. Black Point, Nova Scotia, B0J 1B0
and 324 Clare Avenue, Winnipeg, Manitoba, R3L 1S3
www.fernwoodpublishing.ca

Fernwood Publishing Company Limited gratefully acknowledges the financial support of the Government of Canada through the Book Publishing Industry Development Program (BPDIP), the Canada Council for the Arts and the Nova Scotia Department of Tourism and Culture for our publishing program.

Canadian Heritage Patrimoine canadien
The Canada Council for the Arts / Le Conseil des Arts du Canada
NOVA SCOTIA Tourism and Culture

Sections of this book have previously appeared in the following publications:

Jim Harding, Presentation on behalf of Warman and District Concerned Citizens to the 1980 Federal refinery hearings, and also published as "The Public Health Hazards of the Nuclear Industry" in *Alternatives*, Winter 1981.

Jim Harding, "Spilling the Beans on the Key Lake Spill," *Nuclear Free Press*, Spring, 1984.

Jim Harding, "Provincial NDP Convention Overturns Pro-Uranium Policy," *Nuclear Free Press*, Winter, 1984.

Jim Harding, "The eye of the uranium controversy," *Briarpatch*, April 1985.

Jim Harding and Roy Lloyd, "Free Trade: Pro and Con," *Regina Leader Post*, Nov. 18, 1988.

Jim Harding, "Nuclear Power and Sustainable Development", *Briarpatch*, May 1991.

Jim Harding, "AECL and TRW:The Weapons Connection," *Briarpatch*, May 1992.

Jim Harding, *Aboriginal Rights and Government Wrongs: Uranium Mining and Neo-colonialism in Northern Saskatchewan*, In The Public Interest: Working Paper No. 1, Prairie Justice Research, University of Regina, 1992.

J. Harding, "NDP and Energy: Shades of 1982?" Regina *Leader Post*, Dec. 24, 2002, B2, and Straightgoods news service.

Jim Harding, "Our Deadly Secret: Saskatchewan's Role in Proliferation of MWD," In Y.N. Kly (editor), *The Regina Seminar on the Elimination of Weapons of Mass Destruction*. (Atlanta, Clarity Press, 2006), pp 47-60.

Library and Archives Canada Cataloguing in Publication

Harding, Jim, 1941-
Canada's deadly secret : Saskatchewan uranium and the global nuclear system / Jim Harding.

Includes bibliographical references.
ISBN 978-1-55266-226-7

1. Uranium mines and mining--Saskatchewan. 2. Nuclear industry.
3. Nuclear nonproliferation. I. Title.

HD9539.U72C3 2006 333.8'54932097124 C2006-906395-8

Dedication

Bill Harding, 1911–1993
father, colleague, friend and loved one all at once
(Photo taken at an anti-nuke rally, Saskatoon Bandshell, 1979)

CONTENTS

FOREWORD
by Helen Caldicott

Canada enjoys an excellent global reputation for its past support of multilateralism, international peacekeeping and the United Nations. In this vein, while facing intense anti-war public opinion, Canada's Liberal government decided to stay out of George Bush and Tony Blair's illegal war on Iraq.

However, Canada, like Australia, is hypocritical in its approach to international peacekeeping and nuclear issues. As this book shows, Canada has provided the fuel for U.S. and British nuclear weapons since the beginning of the nuclear age, and it has also provided uranium fuel for nuclear reactors and power plants around the world, in particular the Candu reactors, which use un-enriched uranium and which produce plutonium suitable for weapons' production. In that guise Canada has been directly responsible for lateral nuclear proliferation in countries such as India.

Canada is also a very active member of NATO, which shares facilities with the U.S. to initiate a nuclear war between Russia and the U.S., to coordinate the misguided U.S. missile defence program and to actively participate in the impending U.S. weaponization of space. Of the 30,000 nuclear weapons in the world today, Russia and the United States own 97% of them, many of which are on "hair-trigger alert," ready at any time of any day for instantaneous launch. Such a catastrophe would not only kill millions to billions of people, but it would also induce a nuclear winter with catastrophic climatic events, much worse than the current predicted levels of global warming.

The author of *Canada's Deadly Secret*, Jim Harding, a retired professor of environmental and justice studies and long-term peace and environmental activist, unveils the dark side of nuclear politics in his home province, which bears the distinction of being the largest uranium-producing region in the world, and he challenges us to explore how Canada has consistently been complicit and instrumental in the expansion of the global nuclear system. The book targets one Canadian province — the prairie province of Saskatchewan — to chronicle Canada's long history of involvement in uranium mining. This book is a detailed case study of the front-end of the world's nuclear system — uranium mining.

Harding takes us back to the 1950s and the major role that Saskatchewan played in the first nuclear arms race between Russia and the U.S. He harks back to the 1970s, when provincial public inquiries in Saskatchewan were conveniently used to legitimize the expansion of uranium mining. He moves on to the 1990s to show how the provincial government ignored warnings of environmental reviews that were devastating to uranium mining.

Harding shows how the profitable high-grade uranium in Saskatchewan's north, like uranium mining in Australia, involved the devastation of Aboriginal rights, and how government and private corporations cajoled and bribed northerners facing severe unemployment and poverty with "jobs at any cost." Harding documents the terrible shortfall of promised opportunities, an ever-deepening structural underdevelopment in the north, and the long-term threats to human health of the uranium miners and the people who lived and live adjacent to the mine and their tailings.

It may surprise some readers that a so-called enlightened social democratic government — Saskatchewan's New Democratic Party — spearheaded this wickedness. Accepting at face value the tenets and short-term benefits of globalization, the reigning social democrats used the public purse and Crown corporations to launch the "uranium boom." Shockingly, the Crowns were later privatized to become one of the largest uranium multinationals in the world, Cameco, which now owns nuclear plants elsewhere in Canada and is active in uranium mining in several countries including Australia.

Harding exposes the role the government played in perpetrating nuclear propaganda through the disinformation campaigns of its covert Uranium Secretariat and penetration of the public education curriculum. Such immoral behaviour on the part of the nuclear industry is ubiquitous. The U.S. and British nuclear industries have spent hundreds of millions of dollars in an intensive and fallacious propaganda campaign to convince politicians and the general public that nuclear power is the answer to global warming. The opposite is true, for in addition to producing vast radioactive wastes, the nuclear fuel chain uses massive amounts of fossil fuel and produces large quantities of global warming gases in its own right.

Meanwhile the government of Saskatchewan has stupidly embraced global and continental non-renewable energy markets, while failing to introduce even rudimentary conservation or sustainable energy policies at home. Under the rule of what Harding calls the Nuclear Development Party (NDP), the province has earned a reputation of not only being the world's largest exporter of uranium but of producing the country's largest per capita greenhouse gases. Harding argues that the disastrous impacts of this irresponsible policy of uranium expansion may soon overshadow Saskatchewan's positive reputation of being the birthplace of Canada's public healthcare (Medicare) system.

But this is no parochial book about one environmental struggle in one place. It is a seminal book that shows how interlocked Saskatchewan's environmental and policy crisis is with global threats and challenges. *Canada's Deadly Secret* exposes not just the "environmental ticket" being used by the nuclear industry to try a comeback. He also explores the deadly corporate planning processes that reveal the growing partnership between the oil and nuclear industries. The proposed twinning of nuclear power to help extract the greenhouse-gas-laden heavy oil in western Canada's tar sands is quite extraordinary.

Harding convincingly argues that sustainable development must be non-nuclear. He would like to see proportionate representation to help create a political base for Green politics to help push to end the billion dollars of subsidies to the Canadian nuclear industry.

In summary, Harding shows the central role of his home province and Canada in the growing planetary threat from nuclear contamination and proliferation. During its half century of uranium mining, as well as producing uranium for U.S. and U.K. nuclear arsenals, it has continued to provide the depleted uranium (DU) currently being used by the U.S. military for the production of "penetrating" nuclear weapons and for deadly DU weapons, which are contaminating the Middle East with aerosolized uranium micro-particles. These enter human lungs and guts, causing epidemics of cancer and severe congenital malformations in Basra and other Iraqi precincts. This Canadian uranium 238, which contaminates the cradle of civilization, has a half life of 4.5 billion years.

Harding rues this opportunistic amorality that is catching up with his province. Though it has no nuclear power plants, Saskatchewan is now one of the areas targeted for nuclear waste storage, as the increasingly integrated uranium/nuclear industry tries to use his compromised province as a site for expanding the full nuclear fuel system.

The next few years seem destined to bring the non-nuclear movement back to life in this most strategic area of the global nuclear industry. It is therefore imperative that the revelations from this meticulous nuclear chronicle nudge the Canadian and Saskatchewan public from their collective nuclear amnesia — to become suitably mobilized to save this planet.

Helen Caldicott, MD
President, Nuclear Policy Research Institute
Washington, DC

Preface

TELLING THE STORY AS IT UNFOLDED

I likely had to write this book. The struggle to create sustainable options to the nuclear industry is so intertwined with my family and political life that the book is somewhat of an autobiography. When I was president of the Saskatchewan New Democratic Party (NDP) Youth in the early 1960s I still naively believed that the secular social gospel of the Cooperative Commonwealth Federation (CCF) and the NDP was the way to peace and social progress. Tommy Douglas had been one of my childhood heroes. After winning a teenage safe driving award in the late 1950s I used to drive Tommy to and back from far-away political rallies to help conserve his energy. When I found out in the late 1970s about the complicity of his Saskatchewan CCF government in the nuclear arms race I had a huge identity crisis.

My passion for a non-nuclear society began in the late 1950s with my involvement in the Combined Universities Campaign for Nuclear Disarmament and the Montreal-based journal *Our Generation*. This passion continued, as I became an anti-Vietnam War and student power activist in such groups as the Student Union for Peace Action and Students for a Democratic University in the 1960s. Ironically, at the time of such intense anti-nuclear war activism I knew next to nothing about uranium mining in my home province. I was delighted when Premier Tommy Douglas agreed to speak at one of our large "Ban the Bomb" rallies at the Legislature in the capital city of Regina.

I became more aware of the hazards of nuclear power while teaching in environmental studies and being around Ralph Nader-inspired activists who started the Ontario Public Interest Research Group at the University of Waterloo in the early 1970s. After moving back to my home province in the late 1970s, as a founding member of the Regina Group for a Non-Nuclear Society (RGNNS), I helped publish several small books on the non-nuclear option, including the 1978 provincial bestseller *Correspondence with the Premier*. In this book my now-deceased father, Bill Harding, challenged then NDP Premier Blakeney to rethink his righteous and simplistic assumptions about expanding the uranium industry in the north. As a RGNNS researcher-activist I became aware of the hitherto hidden connection between Saskatchewan uranium and nuclear weapons going back to the Cold War of the 1950s and 1960s.

One thing led to another, as happens in such civil society politics. My involvement in the non-nuclear movement grew to have a global and intercultural reach. On the first anniversary of the Chernobyl catastrophe in 1987 I spoke to Green European parliamentarians about the global dangers of Saskatchewan's uranium mining industry. Because of the networks my spouse Janet and I formed on this eye-opening trip, we launched and helped coordinate the program for the International Uranium Congress (IUC)held in Saskatoon in June 1988. This was endorsed by 160 environmental organizations from twenty-two countries and brought over 200 delegates from around the globe to the city that headquarters some of the world's largest uranium multinationals.

And more networks formed. I was asked to act as a resource for the Keewatin Regional Council when the Inuit in the Baker Lake region of the Eastern Artic were confronted with and ultimately stopped a uranium mine. Due to this work I participated on the Board of Listeners at the World Uranium Hearings held in Saltzburg, Austria, in 1992. The global scope of nuclear devastation really sunk in; the stories told in the presence of 500 delegates from every corner of the planet were so overwhelming they were almost immobilizing. When I had a chance to attend a workshop for personal learning, I chose to sit with Joanna Macy in her session on Despair and Empowerment. This book is very much an externalization of the awareness-creating despair that I have carried since that time.

In the early 1990s we were bombarded with the most concerted campaign of the nuclear industry as it desperately tried to expand nuclear power into Saskatchewan. With the non-nuclear movement waning and disillusionment setting in, I struggled to keep up my activism and research practice. In 1994 my son Joel and I visited friends in Australia and contacted some anti-nuclear groups and Aboriginal land councils. By then I had realized that Australia would become the next major front-end of the global nuclear system after the Saskatchewan uranium boom peaked.

I have always tried to push my activism well beyond "preaching to the converted" and have strenuously engaged with people within pro-nuclear interest groups. In 1993 I was invited to speak to professional and technical personnel in the nuclear industry at the Canadian Nuclear Association (CNA) annual meeting in Saskatoon. Many of my anti-nuclear colleagues were outside peacefully protesting what the CNA stood for while I spoke inside. I have learned much of what is in this book from such encounters. Sitting in the only bar in the only hotel in Baker Lake, after the March 1989 workshop on a proposed uranium mine, I listened to an employee of a Japanese uranium exploration firm tell me that his daughter would agree with what I had said in the public session, but he needed to keep his job.

Without being involved in activist groups such as the RGNNS (1977–81), the emerging Greens (1981–86) and the IUC (1987–93) I would never have adequately understood the workings of the colonialist, corporate and "scientistic" propaganda at the heart of the nuclear struggle. Without my earlier experiences in Saskatchewan's north, where in the 1960s I did field research in community psychology and helped coordinate the Student Neestow Partnership Project, committed to Indigenous self-determination, I would not have felt comfortable embarking on a journey with such vast cross-cultural dimensions. Without involvement in such alliance-building actions as the 1980 American Indian Movement inspired Caravan for Survival, which travelled from Regina to La Ronge, Saskatchewan, to protest uranium mining, I wouldn't have become as aware of the intricacies of post-colonial politics. My international activism allowed me to discern the global dimensions of the nuclear military-industrial system, and I was able to examine all this through the lens of an interdisciplinary education in the sociology of knowledge and the ideology and logic of scientism.[1]

It should therefore come as no surprise that the nuclear controversy became an intimate part of my academic research. As director of research for Prairie Justice Research at the University of Regina in the 1980s I headed up the three-year Uranium Inquiries Project, funded by the Human Context of Science and Technology division of the Social Sciences and Humanities Research Council of Canada. This gave me the chance to explore the pro- and anti-nuclear worldviews in some depth. I published and presented widely in academic as well as in mainstream journals and conferences. I presented to nearly every provincial and federal hearing on uranium mining, refining and nuclear wastes held in Saskatchewan since the late 1970s and, because of this, was asked to act as consultant to the National Film Board in the production of its award-winning 1990 film *Uranium*.

In this book I chronicle these thirty years of interrelated activism and research within the collective struggle for a sustainable non-nuclear society. I hope my voice — as a research-activist, prairie cooperativist and adult educator — helps in this time of ecological brinkmanship. Other vital stories still need to be told of how Indigenous people, concerned about inherent rights and forced underdevelopment, have approached the expansion of uranium mining. Saskatchewan's longest-acting non-nuclear group, the Saskatoon based Inter-Church Uranium Committee, needs to tell its story of unflinching ecumenical witness, in the community and in the courts, to try to stop the corporate assault on creation. And certainly groups based outside Saskatchewan have their important stories to tell — groups such as Greenpeace, who initiated an important campaign in the 1980s on Saskatchewan's Cigar Lake uranium mine, and the European Greens, who showed solidarity with the Inuit in Baker Lake. All voices need to rise up in protest, and in celebration of our past and future victories.

The book is ordered in time as the issues and challenges arose and thus reflects the narrative of my own deepening awareness and understanding.

The reader is invited to become a participant observer in this evolution, travelling with me and other activists who did not quite know at the time what was coming next. Other than updating the names of companies and organizations that have changed, the chapters emphasize what we knew then — information that was also available to the nuclear proponents. This is not a pat, after-the-fact, academic analysis. Following events as they happened, the unfolding story shows how layers of issues and understanding became interwoven into a coherent evolving non-nuclear perspective.[2] The reader is encouraged to stay with the drama, the successes and frustrations of the struggle as it actually took shape. This narrative will be relevant preparation for the inevitable uncertainties in the continuing struggle for a sustainable society.

Notes

1. Jim Harding, "The Ideology and Logic of Scientism," doctoral dissertation in political science, sociology and anthropology, Simon Fraser University, 1970.
2. The results of research on the paradigm shift occurring with the growth of the non-nuclear movement will be of interest to some readers and social scientists with a historical and post-modern bent. See Jim Harding and Beryl Forgay, *Ten Years after the Uranium Inquiries in Saskatchewan: A Follow-up Study of Participants' Attitudes and Concerns*, In The Public Interest Series, Research Report No. 2, Prairie Justice Research, University of Regina, April 1992.

ACKNOWLEDGEMENTS

There are some without whose help *Deadly Secret* would have remained a pile of forgotten articles and a "big" idea. In 2003 my sister Ruth Pickering began preliminary editing of draft chapters based on writings over three decades and encouraged me to stick with my vision of the book as a story of the unfolding anti-nuclear struggle in Saskatchewan. Had our dad, Bill Harding, not left a small trust fund to continue resourcing for a non-nuclear future, the project would not have been practical. Thanks to the Bill Harding Trust Fund for supporting editorial assistance and a small publishing grant.

And had I not gotten to know Richard Therrien when he edited an earlier book, *After Iraq* (Fernwood, 2003), and had he not agreed with little notice to help me whip the early drafts into a coherent narrative, I probably would have again put this personally staggering project aside. Richard's suggestions about "voice" and reorganization, his skilful editing of difficult material and his steadfast encouragement when the project got tough all shaped the outcome. Though I remain fully responsible for the research and content, what you experience as you read *Deadly Secret* is a result of our creative collaboration.

And a huge thank you to Helen Caldicott for writing the Foreword. Helen continues to challenge and inspire us to take heart and to take meaningful action on the nuclear threat. We are all in deep gratitude for her sticking with this sometimes overwhelming and despairing "issue." Her speaking trip to Regina and Saskatoon in March 2007 has already sparked new initiatives — such as the Non-Nuclear Network.

And, of course, a special thank you to the folks at Fernwood. Wayne Anthony's vision of this as a trade book and his detailed notes on the full manuscript became the basis for a full rewrite that helped "pull" the book together. While I initially resisted the major rework, Wayne's judgment was correct and I am indebted for his frank feedback. Also a note of gratitude to John van der Woude for designing the book cover. And it's again been a pleasure working with Brenda Conroy, who did the copy-editing, Beverley Rach, who coordinated production, and Debbie Mathers, who corrected the final manuscript. Their congenial natures made the last jobs, which really make or break the project, something to look forward to.

Without the ongoing non-nuclear activism of many hundreds of Saskatchewan people there wouldn't be a story to tell. My gratitude to all the "lifers" who let the evolutionary challenges this global issue posed for us in Saskatchewan enter their hearts and souls. I feel especially grateful to the local elders of our struggle, Bill Harding and Stan Rowe (now deceased) and

Masie Shiell and Al Taylor; and to those always helping from afar, especially Gordon Edwards and Rosalie Bertell; and a special mention of the vital work of the Saskatoon-based Inter-Church Uranium Committee, the longest lasting non-nuclear group in Saskatchewan, which has kept our attention on this issue when we have been distracted from the great nuclear threat facing us.

Also my gratitude to other long-term activists: to Anne Coxworth, coordinator of the Saskatchewan Environmental Society, especially for her bringing broader awareness of the potential of energy efficiency through her involvement in the Energy Options Panel; to Stephanie Sydiaha, who keeps coming home to the uranium/nuclear issue, for her invaluable networking in the north and Europe; to Miles Goldstick, now of Sweden, who kept trying, sometimes in vain, to animate local activism to challenge our nuclear amnesia; and to Peter Prebble, long-time Saskatoon NDP MLA, who knows how tough it has been to keep focused on a sustainable future while being in the eye of the nuclear storm. The magnitude of this issue is revealed by the diversity of the practices and commitments shown by these people.

I would like to make special mention of the defunct Regina Group for a Non-Nuclear Society, which in its heyday created a whole larger than its parts, with diverse decentralized committees and 500 people on its mailing list. Particular mention of Jan Knowles and Janet Stoody, the "two Jans of the RGNNS," for their inspirational songfests throughout the province. Some RGNNS activists went on to become the Regina core group that helped organize the International Uranium Congress (IUC), which brought over 200 people from twenty-two countries to Saskatoon in 1988: Jim Elliot, Brian Johnson, Valerie Overend, Clare Powell, Dave Weir and my resilient, multi-talented, heartfelt spouse Janet Stoody. And to Rod (now deceased) and Rose Bishop, Brenda Dubois, John Graham, George Smith, Adele Ratt and others who had the courage to express opposition to the nuclearization of the north in the north; to John Willis, who demonstrated that Greenpeace could indeed work with Saskatchewan activists; to Jack Hicks, who did exemplary work with the Inuit to stop the Baker Lake mine; to Jamie Kneen, now of MineWatch, who picked up in the north where others left off; and, finally, to Ralph Torrie, with whom I had my first in-depth discussion about uranium and nuclear wastes, while on a ferry going north to Manitoulin Island in 1976. And let us not forget Saskatchewan pioneers demonstrating the practical alternatives to the destructive inefficiency of the nuclear industry: to Ken Kelln, Vic Ellis and others steadily building renewable energy options.

The IUC was not viable after its one-man secretariat — my father Bill Harding — became terminally ill in 1992. After that shock and loss my main co-conspirator in non-nuclear community resourcing became Dan Parrot, now working with the Sierra Club, without whose encouragement I might never have written pamphlets, leafleted NDP conventions, appeared at hearings or participated in the public debates that shaped the later chapters of this book.

Without research there can be no effective activism. I wish to thank fellow researchers in the SSHRC-funded Uranium Inquiries Project, housed in the School of Human Justice at the University of Regina after 1983: Ingrid Alesich, Dave Gullickson, Malcolm Harper, Lillian Sanderson and especially Beryl Forgay. Also, many thanks to research coordinators Rae Matonovitch and Mary Gianoli, and research secretaries Linda Bradley and Diane Vandenberghe for their essential contribution. Publications from our collective research[1] provide a backdrop for much of this book and remain one of the best sources of information on the ideological and political dimensions of the nuclear controversy.

A note of thanks to Helen Hawley, a part of our shrinking extended family, who for years sent my father and I clippings from England on the nuclear industry; to those who helped me locate pertinent information as the book's deadline approached: Jim Penna, Bill Adamson and Graham Simpson; to my spouse Janet Stoody and friends Scott Preston and Steve Izma for insightful suggestions on the draft introduction; and to Kairos colleagues, especially Dick Peters, for encouraging me to complete the project.

My final acknowledgement is to all the kind-hearted, steadfast people I met on three continents who have not lost their "great faith, great doubt and great perseverance" through years of difficult anti-nuclear activism. This includes people working with Nuclear Guardianship, Peace Research, Alternative Energy Institutes, the World Uranium Hearings and Green politics.

This has been an arduous project. When I came back to it with uncertain fervour in July 2006, I was still in the midst of deeply grieving my mother, Bea, who died on the March equinox. I really wanted to get back to writing another book, *Between Ages* — about ancestral (family) heritage, personal awakenings and ecological spiritual possibilities in our upside-down world — but this "material" was still far too poignant. Also I somehow knew I had to "clear the decks" and finish *Deadly Secret* to be able to move on as well as prepare myself for the coming battles over nuclear expansion. Janet and I were stunned at my mum's passing and exhausted from years of care-giving. Bea wanted me to complete this book on nuclear politics, which largely comes out of her husband, my father Bill's and my activism, and I regret she — also a writer — never saw it come into being. However, I found the energy for the task in the vacuum left by her passing, and the book is a testament of my devotion to her.

I dedicate this book to my father Bill, who after an exhausting career in international civil service in such places as Somalia and the Philippines returned to his province of birth to provide leadership and inspiration to the emerging development and environment NGOs. He is, among other things, a founding person of the Saskatchewan Council for International Co-operation. *Deadly Secret* is then, in part, a family story. The raising of our three boys parallels these decades of intense environmental activism, with our youngest Dagan turning five in Rome during a European trip on the first anniversary of Chernobyl; our middle son Joel travelling with me after finishing high school to network in northern Australia; and prior to completing his legal training, my oldest son Reece working at the West Coast Environmental Law Foundation. Like my father and me, my sons will be challenged in their own vocations — law, biology and cultural studies — to contribute to creating a sustainable future for their grandchildren's generation. Since it took three generations for the nuclear nightmare to unfold, we should be prepared to take however many generations are required to turn humanity back onto a sustainable, non-nuclear path. *Deadly Secret* is my offering to that ongoing project.

Note

1. See several working papers and research reports, under the series: In The Public Interest, Prairie Justice Research, University of Regina.

Introduction

ONE WORLD, ONE FUTURE

Saskatchewan is currently the largest uranium-producing region in the world. It accounts for about 30% of annual world uranium production[1] and exports uranium to most of the world's major nuclear powers. In September 2004 Saskatchewan's NDP Minister of Industry and Commerce, Eric Cline, giving the keynote address to 500 nuclear delegates at the annual meeting of the World Nuclear Association in London, bragged about this and called for even more production of uranium from his province.[2] In February 2006 Cline and NDP Premier Lorne Calvert met with U.S. Vice President Dick Cheney to secure more U.S. investment in Saskatchewan's uranium and oil industries. This rise to the pinnacle of uranium production, of which the Premier and Minister were so proud, took several decades to develop, beginning with the first uranium mines in the 1950s near Uranium City in Saskatchewan's north.

Polite, peaceful, benign Canada is not a country that readily comes to mind when the topic of the world's nuclear powers comes up, but few realize that the uranium for the bomb that instantly killed over 100,000 people in Hiroshima came from the Port Radium mine in the Northwest Territories. And few would be able to identify the province of Saskatchewan as the initial source of substantial amounts of the depleted uranium (DU) that is now routinely used in modern "conventional" weaponry.

Though hunting and gathering has been largely displaced by agriculture and agribusiness, I consider myself — along with most Saskatchewan people — to be a member of a "land-based people." Born on the experimental farm south of Swift Current (aka Speedy Creek) and now living in my retirement on an ecology preserve in the Qu'Appelle Valley, the land has consistently nurtured and deepened my ecological activism and spirituality. When I enrich the gardens, where my spouse Janet and I live, with our organically grown alfalfa hay or nutrient-rich compost, and later taste the delectable food the cared-for land has brought forth, I sometimes think of uranium mining and its end-uses contaminating lands far away that have produced good food and livelihoods for thousands of years.

I am not only referring to atomic attacks, or atmospheric testing, or reactor accidents, nor is the contamination I am thinking about limited to my home province. In Iraq olive groves dating back 6,000 years have been contaminated with DU bombs, some of which likely have their source in Saskatchewan's north. A week after the Shock and Awe pounding of Baghdad in March 2003, uranium aerosols from DU weaponry found their way into the English atmosphere — and I presume the lungs of Englanders — some 2400 miles away.

Driving through the populated cities and rural areas of southern Saskatchewan there is nothing to indicate that much of the world's uranium began its journey here in this province. The DU in "conventional" weapons, the uranium in stock-piled nuclear bombs, the uranium used in nuclear energy and the millions of tons of highly radioactive mine tailings littering the landscape — these all form part of a single intertwined global nuclear system. And Saskatchewan now holds the distinction of being the world's major front-end supplier of that system.

In the southern part of the province, things look more or less the way they have for decades. Yes, most of the grain elevators are down and gone, and the Big Box stores continue to amass near the spreading suburbs as they do elsewhere on the Prairies. But there are no nuclear power plants, as there are around the Great Lakes near the city of Toronto, nor are there any nuclear missile launching pads to remind us of the bigger nuclear reality, as there are south of us in Montana.

You have to travel to the far north, well past La Ronge, which is mid-way between the U.S. and the Northwest Territories border, to see the ecological scars created by the huge uranium mines at such places as Rabbit, Cluff, Key and Cigar Lakes, or McArthur River. The huge pits and the sheer magnitude of radioactive tailings are heart-stirring to anyone who has eyes to see.

But even the deep scars and radioactive wastes left on the land do not tell the whole story; to fully realize what the uranium industry is doing to the planet, you'd have to trace the yellowcake, shipped from the uranium mills in the north to Cameco's refinery and conversion plant in Ontario or to enrichment plants in the U.S. or France. You'd have to trace its conversion into fuel used in Candu and other nuclear power plants in many countries and its consequent transformation through nuclear fission into long-living toxic nuclear wastes like plutonium, a substance never before seen on planet earth. And, if you've got the stomach for it, you'd have to trace the fuel being converted to the fissionable material that ended up in thousands of nuclear warheads; you'd have to witness the DU left from the enrichment processes going into the uranium bullets now being questioned by the United Nations[3] but used nonetheless by the U.S. or NATO in its four most recent wars, and used in the casings of H-bombs ready to play their part in any number of possible genocides brewing in the war-rooms of nuclear weapons powers. Only then, driving through this Canadian heartland armed with such knowledge as is presented in this book, would you begin to comprehend the full ecological and moral implications of mining uranium in Saskatchewan.

The historical narrative of Saskatchewan nuclear politics begins with the Manhattan Project and the dropping of two atomic bombs on Japan in 1945, after which the U.S. began to build its nuclear arsenal; a few years later the Canadian government used the *War Measures Act* to expropriate Eldorado Gold Mines. In 1946 a new Crown corporation, Eldorado Nuclear, staked its chief uranium claims in the Beaverlodge area of northern Saskatchewan.

In 1952 the U.S. exploded the first hydrogen bomb — and the Saskatchewan uranium boom was under way. After President Eisenhower's "Atoms for Peace" speech in 1953, expectations for uranium-dependent nuclear-generated electricity rose considerably, and from 1962 to 1966 the Canadian government took steps to stabilize the uranium industry, including stockpiling uranium. The nuclear weapons-based industries in both Ontario and Saskatchewan hoped to make the transition to the growing nuclear reactor market. This looked promising; from 1964 to 1967 there were sixty orders for commercial reactors placed from the U.S., and between 1966 and 1970 several orders were placed from Europe, Japan and other countries.

But the price of uranium collapsed in 1971, and from 1972 to 1974 a uranium cartel, which included Canada, was created to keep the price of uranium artificially high. By 1973 the Saskatchewan Blakeney NDP government was looking more seriously at an earlier suggestion to establish a Crown corporation to stabilize uranium development.

In 1973 the U.S. Atomic Energy Commission, holding a monopoly on enrichment facilities in the western world, changed its system of providing enrichment services to long-term fixed commitments. This required reactor companies to place orders by 1974 or risk going without fuel into the 1980s, creating a surge in reactor orders and a new market for uranium fuel.

In 1974 the newly elected Australian Labour Party put uranium mining expansion on hold due to social and environmental impacts and concerns for Aboriginal rights; across the Atlantic political unrest in Gabon and Niger threatened France's uranium supplies, and as a result the French uranium company Amok announced their plans to develop the Cluff Lake mine, rich in uranium deposits, in northern Saskatchewan.

Scientism is the ideology that turns self-critical scientific inquiry into a religion that worships a technological priesthood of corporate "experts" who are out of reach of democratic processes. The propaganda of the nuclear industry, in both its energy and weapons manifestations, relies heavily on such rhetoric and on the collective historical and ecological amnesia of the general public, which sustains it. This propaganda is rendered even easier by the largely invisible effects of nuclear development. Most of us never come into direct awareness of the uranium mines in Saskatchewan's north, and if we do, these are described neutrally in terms of resource development bringing royalties to the province and employment to the region. They are not described as producing tailings or spent fuel for millennia or continuing nuclear weapons proliferation. And without deep curiosity and knowledge about the end uses of uranium mining, it will never be seen as anything other than just another resource industry.

The sole purpose of the pro-nuclear lobby is to create a lucrative short-term market for uranium fuel and Candu reactors, and the global ecological stance taken in this book presents a challenge to their short-sighted self-interests. This narrow parochial economic perspective has been adopted by the Saskatchewan NDP, which has repeatedly used this short-term rationalization to defend their political, bureaucratic and even corporate interests. But this rationalization, couched in the language of economic necessity, is comparable to the amoral rationalizations used in the past to justify the economic benefits of slavery. Or, to use a more contemporary example, to rationalize the economic benefits of exporting illicit drugs, for example, cocaine (from Columbia) or opium (from Afghanistan) to a lucrative market of desperate people abroad. Out of sight, out of mind, with a dollar in hand: this is the psychopathic amorality described in the 2003 book by Joel Bakan *The Corporation*, and film of the same name. But for present and future generations, the export of uranium has far more devastating consequences than illicit drugs; there is really no comparison when you consider the millions of people, worldwide, already victimized by the nuclear industry and the millions more for generations to come.

A conversion strategy, not political correctness, is required for moral-ecological awareness to grow. This strategy involves a phasing-out of nuclear power and uranium mining and a shifting of society's resources to conservation and sustainable energy systems, which can actually prevent further pollution, radioactive contaminants included, and reduce the impacts of global warming. Saskatchewan could now be a leader in this had it not stubbornly stuck to the nuclear path.

Canada's Deadly Secret is a testament to the creation of transformative political ecological consciousness out of an ongoing collective engagement. In giving shape to this story, I was startled — after three decades of activism — by the magnitude of the resources that the uranium and nuclear industries had thrown at the province. This includes the full weight of the seemingly bottomless corporate coffers and the propaganda machine of the entire nuclear system, from extraction to end-use; both federal and provincial states with the Crown corporations and multinationals they created; and the near-monopolistic media that uncritically supported, and still support — even worship — economic growth and energy consumption at any cost.

But perhaps more importantly I was struck by the fact that those enormously powerful and manipulative conglomerates, acting with government and media collusion, have gained so little ground. The effectiveness of our clumsy and hesitant resistance to their unholy plans is impressive.

The intention of this book is to connect the dots between Saskatchewan's uranium mining and the other vital issues along the nuclear industry's global pathways. I hope it will bring historical context to the continued necessary resistance to the government of the day — the NDP or its replacement — as it seeks a quick fix to its economic struggles, obscures the broader threat of global warming and continues to ignore the challenge of sustainable development, by laying plans to expand the nuclear industry in Saskatchewan.

Notes

1. See Saskatchewan Eco Network <http.econet.sk.ca/issues/mining/index.html> accessed July 20, 2007.
2. "Saskatchewan's Uranium Resources Promoted in London," News Release, Saskatchewan Executive Council, Sept. 10, 2004.
3. Leuren Moret, "Depleted Uranium is WMD," *Battle Creek Enquirer*, August 9, 2005, available at <http://www.commondreams.org/views05/0809-33.htm> accessed June 2007.

chapter 1

SIDE-STEPPING ABORIGINAL RIGHTS

When I returned to my birth province in early 1977 the uranium issue was ramping up. Many New Democratic Party (NDP) delegates to the fall 1976 provincial convention had been shocked and furious when they found that their social democratic government was deeply immersed in the uranium business. By then the Saskatchewan Mining and Development Corporation (SMDC), created by the Blakeney government in 1974, was for all practical purposes "The Uranium Corporation." It had already laid the groundwork for the development of a huge uranium mine at Cluff Lake. To avoid grassroots pressure for a full scale moratorium on this venture and the public perception of a serious fracture within the party, convention delegates adopted a compromise motion to refer the proposed uranium mine to a public inquiry. The Blakeney government had no realistic political option but to follow suit.

In 1977 the Cluff Lake Board of Inquiry (CLBI), headed by Justice Edward Bayda, held meetings throughout Saskatchewan. Briefs were presented by the Saskatchewan Environmental Society, United Church and many other organizations and individuals. I presented one for the Regina Group for a Non-Nuclear Society (RGNNS), and I also presented research from my work at Saskatchewan Health, where I was director in the addictions research branch.

The uranium industry was making grandiose claims about all the positive socio-economic impacts their new mines would bring to northerners, but the government wouldn't allow any time for serious research into their claims. I knew from my social science background that capital-intensive, boom-bust mining ventures brought many negative impacts. The civil service seemed muzzled about raising any critical questions, perhaps because Premier Blakeney was already on record saying the government already had a policy as there were existing uranium mines. In its brief, the Department of Northern Saskatchewan begged the whole question by saying uranium mining "will certainly not create any problems that haven't been experienced in the past," by which it didn't mean to admit that there were far more than expected lung cancers among former uranium miners. I was in a bit of a bind as a probationary civil servant directing a research unit that had findings with implications for the controversy over social impact, as ministers and their deputies weren't soliciting any of this. I finally decided to appear unofficially and present some of our findings on the north.

We had just done an analysis of alcohol-related disabilities, comparing very isolated communities with frontier communities that were on northern roads used for resource extraction. We found an 11:1 ratio of *Liquor Act* violations in the Indigenous frontier communities of La Loche, Buffalo Narrows and Ile a la Cross compared to the mostly non-Indigenous uranium mining towns of Uranium City and Eldorado. But we had also found a 6:1 ratio when we compared these three Indigenous frontier communities with the isolated ones of Stoney Rapids and Black Lake, and I thought the startling contrast should be brought to the attention of the CLBI.[1]

When the Board of Inquiry referred to our research in their final report they ignored the results and simply repeated their mantra that new uranium mines were the economic promise for the north. They never took a systematic look at the matter of social impact and refused a time extension that would have allowed them to do so. When a content analysis of the CLBI transcripts confirmed that all panel members were unconditionally pro-nuclear, we understood why.[2]

Not surprisingly the Board applied a narrow interpretation to its terms of reference, notably failing to investigate the implications of uranium mining for Aboriginal rights. With the government's blessing, the Inquiry ignored the call by all Indigenous organizations — Métis as well as First Nations — for a moratorium until numerous land claims could be settled. Such a moratorium was consistent with a wide body of law on Aboriginal rights even before the 1982 Charter.

Participation in the Inquiry by Indigenous people was unexpectedly high, and the message was loud and clear: the vast majority of the people to be most profoundly affected by the huge new uranium mine at Cluff Lake either supported their leaders' call for a moratorium or were opposed to uranium mining going ahead under any circumstance.

Though the CLBI refused to consider the threat uranium mining posed to Aboriginal rights, it was to eventually recommend uranium revenue sharing with the north and a Northern Development Board, as a means to ensure "fair" distribution of the economic benefits and to ameliorate the social costs of uranium mining. But though the provincial government had earlier left the impression it would consider revenue sharing, it rejected both proposals, using its own calculations to argue that more was already being spent in the north than uranium revenues would provide. Seeing no way to stop the uranium bulldozer, some Indigenous leaders began to look for ways to get trickle-down benefits from the uranium go-ahead. The processes and policies used by the NDP government in attempting to subdue the "anti-nukes" in the party not only publicly legitimized the expansion of uranium mining and undermined public participation, they also perpetuated neo-colonialist actions against the Indigenous people.

In the CLBI's 300-page report there is only a half-page reference to Aboriginal rights. It reads:

> Our terms of reference are not sufficiently broad to permit a thorough investigation of that issue and indeed, to have made the issue a part of the present Inquiry would have been a mistake, for, the very nature of the issues dictates that if there is going to be an investigation at all, it should be the subject of a separate investigation.[3]

This exclusion does not stand up to historical scrutiny. With similar mandates, both the McKenzie Valley (Berger) and Alaska Highway inquiries included Aboriginal rights in their investigations. The former recommended a ten-year moratorium and the latter a four-year postponement of a pipeline in order that Indigenous land claims could be addressed. The Churchill River Inquiry in Saskatchewan also discussed Aboriginal rights.

Writing in the *Saskatchewan Law Review*, Law Professor R.H. Bartlett said:

> The Board is clearly empowered to review and recommend conditions regulating the social and economic impact of the project. Such a review necessarily entails a study of the rights in law of the Indian and Native people of Northern Saskatchewan, who represent a significant element in the social and economic structure of the region… the CLBI was remiss, and in error, in construing its terms of reference so as to deny consideration of what Justice Berger termed "the urgent claims of northern Native people."[4]

Indigenous people are actually more than a "significant element… in the region." At the time Indigenous people made up 19,000 of the 25,000 northerners. About 10,000 of these people were Métis and non-status Indians. The remainder were status Indians. If the three mining and northern administrative centres of Uranium City, Creighton and La Ronge are excluded, Indigenous people constituted over 90% of northerners.

Attendance and participation in the CLBI clearly showed that uranium mining was a vital matter to northern Indigenous people. Overall, 165 people, or more than half (57%) of all those who attended the twenty-three local hearings throughout the province, were from the north, as were half of all those who spoke at the local hearings. Of these northerners, 75% in attendance and 71% of the speakers were Indigenous people. Most of those who spoke were Indigenous men, though Indigenous women were better represented than non-Indigenous women. Indigenous people were the vast majority in attendance at all local hearings except the one held in Uranium City, where most of the 60% of the non-Indigenous attendees were in one way or another associated with the uranium industry.[5]

Neither the CLBI nor the NDP provincial government would confront the issue of Aboriginal title and resource development. As "occupants" of the land for centuries the Indians and their offspring were granted certain right to lands and to compensation for any extinguishment of such rights. Bartlett thus writes:

> The people of Indian ancestry of Northern Saskatchewan are accordingly suggested as beneficiaries of the Indian title therein, to be entitled to the mineral resources of that region unless it is considered that such title has been extinguished.

Treaties 6, 8 and 10, passed between 1876 and 1906, have a bearing upon the ancestors of the Plains, Wood Cree and Chippewyan Indians of the region affected by uranium mining. These treaties influence several northern bands; however, Treaty 10 is particularly relevant since it involved the bands most directly affected (Lac La Hache, Stony Rapids and Fond du Lac), as well as Key Lake, where the largest uranium mine in the world was built.

Cluff Lake Inquiry: If Not a Moratorium Then a Boycott

Aboriginal rights were of such importance to Indigenous people that all their organizations in the north argued there should be a moratorium on uranium mining until land claims were settled. If there wasn't going to be such a moratorium, they argued, the CLBI should be boycotted. In a unanimous statement on October 7, 1977, the Meadow Lake and Prince Albert District Chiefs of the Federation of Saskatchewan Indians (FSI) made their views about land entitlements clear:

> [The] Inquiry is not dealing with the question of whether or not uranium mining should expand in northern Saskatchewan but simply the question of how.... To participate in an inquiry having such limited terms of reference would mean that Indian people and their governments (Chiefs and Councils) have already concluded that further uranium development is an acceptable and desirable course of action for the Indian people of northern Saskatchewan. This is not the case.[6]

The northern chiefs viewed boards of inquiry as an instrument to "create an illusion of objectivity and public participation" and they had "no desire to lend further support to such an illusion...." Their legitimate participation would be impossible,

> given the totally inadequate timeframe of the inquiry and the almost

> complete absence of resources to investigate the variety of serious issues and questions related to the uranium industry. To enter an arena where the tremendous resources of industry and government are so blatantly stacked against us, would be politically irresponsible for us as Indian leaders.

Seventeen other stakeholder groups refused to participate in the CLBI formal hearings for many of these same reasons.

With the issues concerning the environment and the social and economic impacts of further uranium mining on the Indigenous people of northern Saskatchewan left unresolved, the northern chiefs stated that, regardless of future considerations, no further uranium development was acceptable until:

a) land selection by Bands with unfulfilled Treaty land entitlement is completed;
b) the Treaty Rights of Hunting, Fishing, Trapping and Gathering are guaranteed against violation;
c) the Treaty Rights for health, economic development and resources management are assured; and
d) we have the time and resources to carefully examine the many serious questions related to the uranium industry.

It is worth noting that the northern chiefs felt compelled to reiterate their position even though the federal government had announced in a 1977 communiqué on land claims that the federal, provincial and Indian governments had come to an "official agreement on the means of fulfilling outstanding Treaty land entitlements of Bands" in Saskatchewan. Under the provisions of the relevant 1871 to 1906 treaties and the Natural Resources Transfer Agreements of 1930 between the federal government and the Prairie Provinces, every Saskatchewan Indian band was to receive whatever land was required to bring the total land per registered person as of December 31, 1976, to 128 acres. The refusal of the northern chiefs to participate in the CLBI suggests that they remained skeptical until the land agreements were finalized.

When the federal communiqué was issued, the FSI had only advanced land entitlement claims for fifteen of twenty-five northern bands. Though agreement in principle had been reached for some bands, the northern chiefs stood firm on the prerequisite of full settlement of land entitlements prior to any consideration of expanding uranium mining.

The communiqué also stated: "Saskatchewan is also prepared to fulfill entitlements to the Bands concerned by providing, instead of lands, opportunities to Bands for revenue sharing in resource development or participation in joint ventures." This suggests that the provincial government may have been considering royalty sharing as an alternative to returning lands to the bands, though once it had the Cluff Lake mine up and going, it rejected the CLBI recommendation of uranium royalty sharing out of hand.

Revenue sharing was proposed because the treaties had not extinguished mineral rights. Some treaty Indians have argued in the case of Carswell Lake, an acknowledged Indigenous archaeological site near the Cluff lake mine, that Treaty 8 did not extinguish title to the land because no land settlement has yet been made. Furthermore the authenticity of the Indian signatures on the original treaty has been questioned. According to Bartlett, writing about René Fumoleau, an Oblate priest who had critically researched Canada's treaties, "on the Treaty 8 documents nearly all of the marks next to the Chiefs' names are identical, perfectly regular with a similar slant, evidently made by one person."

The firm position on uranium mining taken by northern chiefs during the CLBI has already begun to disappear from political memory. Because the government has clearly not been willing to deal with Aboriginal rights in good faith, Indigenous people have been forced into a "take it or leave it" predicament. Suspecting a *fait accompli*, some of the FSI leadership decided to make the best of a bad situation and cash in on uranium mining through such things as trucking and security guard contracts and job quotas at the mines.

Métis Scrip in the North Ignored

The position of the Association of Métis and Non-status Indians of Saskatchewan (AMNSIS) was even more militant. They asserted their Aboriginal rights directly to the CLBI:

> Our people are the Aboriginal inhabitants of the Prairie Provinces, and as an Aboriginal people we have an Aboriginal claim to the land, a claim which is guaranteed in British law in the British North America Act, a claim that is further reiterated in the laws of Manitoba, Saskatchewan and Alberta in the Canada Lands Transfer Act of 1932. These laws have been completely ignored by successive federal and provincial governments. We have been driven from our land in contravention of the laws of Great Britain, Canada and Saskatchewan. In short, our land has been stolen. This is an incontestable fact.[7]

After outlining the severe social and economic problems of Indigenous people who had been denied their Aboriginal rights, the AMNSIS spokesperson concluded that "it is only just that it be our people who determine whether or not this development be allowed to proceed." There was little doubt about how the AMNSIS viewed the proposed uranium mine:

> The proposed uranium development represents only one of hundreds of corporate and government decisions to commit robbery, theft and even genocide against our people. If these comments seem harsh, you know, I think we could substantiate a lot of what we are saying today.

Such substantiation would reach back at least a century, to the failure of the Canadian government to honestly provide Métis scrip for land, as legislated under the *Manitoba Act* of 1870.[8] Bartlett comments: "The Métis have long assented that scrip was provided in circumstances of fraud and manipulation which enabled banks, other financial institutions, land companies, lawyers and small speculators to claim the Métis entitlement." As Métis lawyer Clem Chartier wrote:

> While treaties with the Indians set apart communal tracts of land and recognized other rights, the scrip issued to half-breeds was a specific amount of land which was fully alienable. In addition, by this method of unilateral dealing, the government of Canada also purported to extinguish all Aboriginal title rights possessed by the Métis, including the right to hunt. As a consequence of this imposed scrip system, most of the land fell into the hands of speculators.[9]

About northern Saskatchewan, Chartier wrote that,

> Because scrip could only be applied against surveyed land, a significant number of Métis were immediately at a disadvantage. For example, in the 1906 Treaty 10 area of northern Saskatchewan, 60 percent of the scrip issued was land scrip. To this day there is virtually no surveyed land in that area. As a consequence, the Métis of northern Saskatchewan were deprived of their land base and their opportunity to acquire ownership of land.

Many Canadians do not yet realize that the Métis were the vast majority (80%) of the population in the area that became Manitoba and that the Riel uprisings of 1869 and 1885 were central to bringing the Prairie Provinces into Canadian federalism. The evidence is now overwhelming that the federal government planned to cheat the Métis and undermine their Aboriginal rights. After the 1869 Fort Gary uprising, Prime Minister John A. McDonald stated: "These impulsive half-breeds have got spoiled by their emeute (uprising) and must be kept down by a strong hand until they are swamped by the influx of settlers."[10]

The outright hostility towards Aboriginal rights expressed by McDonald continued into the next century and was reflected in the 1977 CLBI. Witness the cross-examination of AMNSIS spokesperson by Derril McLeod, the lawyer of the French company Amok, which was developing the Cluff Lake mine, who sidetracked the issue by challenging the AMNSIS witnesses to prove they were legitimate northerners. When their Indigenous and northern roots were indicated he commented, "Well, I consider myself a northerner too, but I didn't come from quite that far north." Judge Bayda, who presided over the CLBI, participated in this paternalism.[11]

This shift to a facetious and petty discussion of individual geographic origins showed that the proponent Amok and the CLBI were not willing to respond seriously to the AMNSIS concern for collective Aboriginal rights. This use of the geographic definition of northerner to ridicule legitimate Aboriginal claims is an extension of the well-documented tactic used to colonize the First Nations and Métis of Canada. It was most strikingly demonstrated when the rights of Manitoba's Métis were undermined through the concerted and deliberate European settlement of targeted areas—a sort of low-level genocide by displacement.[12] Many northern Saskatchewan people are the descendants of Métis driven to the northwest by this historical colonization, and from all reliable estimates they will continue to be the majority of the population in northern Saskatchewan. Discrediting the land rights of Indigenous people while accelerating the government-assisted concentration of corporate wealth and non-Indigenous communities within traditional Indigenous territories is reminiscent of the Israeli settlements on Palestinian lands.

The president of the AMNSIS reiterated their position soon after the CLBI report was released. As reported in the July 1978 issue of the Métis magazine *New Breed,* Jim Sinclair announced:

> AMNSIS is not surprised by the results of the Bayda Inquiry — the decision to develop uranium in the Cluff Lake area was made long before the inquiry started.... The more important issue (of Aboriginal rights) must be settled before the Native people can be freed from government dependency and control.

Furthermore, the AMNSIS could not support the proposed Northern Development Board unless it

> had the authority and resources to deal with the...protection of Native rights. If... Department of Northern Saskatchewan officials hold true to their past record, the Northern Development Board will be a useless and powerless board established simply to appease the provincial government and southern non-Natives. It would simply

> give the appearance that Natives have a say in the development of Northern Saskatchewan, when in reality they do not.

Métis and non-status Indians are clearly more vulnerable than band Indians regarding just land entitlement. For the uranium mining industry to expand before the Métis and non-status Indians had the opportunity to establish their legal claim to the land further jeopardized their long struggle for self determination. The unwillingness of the government to deal directly with Aboriginal rights through the CLBI, even though these rights would be directly affected by uranium mining, exposed the lack of good faith in their bargaining over land claims in other forums.

Government Bargaining in Bad Faith

There is a point at which this bargaining in bad faith comes very close to acting outside the law. A federal Cabinet memorandum prior to the sitting of the CLBI shows the government of Canada considered the right of the Métis and non-status Indians to make land claims to be legitimate.[13] The document acknowledges the need of Métis and non-status Indians for self-determination, stating that "non-status Indians and Métis may have legal claims against the federal government and some provinces, and this might be tested in the courts at any time." Later the document explicitly accepts "the prima facie evidence that there exists a class of indigenous people outside the Indian Act that may have justifiable claims to 'Aboriginal title'...." The document states the government's objective "to settle outstanding valid claims, based on Aboriginal title, by negotiation, taking full account of indigenous requirements in terms of land and ecology to sustain a traditional lifestyle...." It recommends that "the government agree to provide funding, on a mutually acceptable basis, to non-status Indian and Métis organizations at once to research legal claims" and that "there is an urgent need for action, especially in relation to the funding of research into legal claims."

The Cabinet document was in part politically motivated, suggesting that government "take a low-key approach in public to avoid an Indigenous backlash like that against the 1969 policy paper" and that government should "continue to work through, and foster Indigenous associations and their moderate leadership." In particular, it stressed the need to give "Native socio-economic problems in Western Cities (and Western Northlands) and in various rural areas... urgent attention to forestall social unrest." Though it may have been at least in part cynically motivated, the Cabinet document did indicate legitimacy for the claims of Métis and non-status Indians to Aboriginal rights.

The biographer of Allan Blakeney, premier of Saskatchewan during this expansion of uranium mining, notes that Blakeney was forthright in his rejection of Métis claims to Aboriginal rights.[14] It is somewhat ironic, with the Métis now included along with Inuit and First Nations under constitutional discussions of Aboriginal rights, that in 1991 Allan Blakeney was appointed to the Federal Commission on Aboriginal affairs.

The commitment to Aboriginal rights is pervasive in the north, and even organizations such as the Northern Municipal Council (NMC), otherwise integrated into the political and administrative structure of the dominant society, affirmed the recognition of their claims. In its submission to the first phase of the CLBI, the NMC stressed the priority of Aboriginal land rights. In addition to affirming their own rights as non-treaty northern Indigenous people, members of the NMC indicated their solidarity with the land rights of treaty Indians in the north.

With Aboriginal rights ruled out by the inquiry, it is easy to understand how some members of the Indigenous communities viewed their dilemma as a "take it or leave it" option. All interest groups, including the CLBI itself, advanced their own self-interested interpretation of how "northerners" viewed uranium mining. Pro-uranium groups implied that northern Indigenous people supported the Cluff Lake mine because they would be able to personally benefit from its opportunities. But a three-year study involving sampling, coding and content analysis of inquiry participants' attitudes and viewpoints provides a much more credible picture.[15] *None of the Indigenous participants (in the sample) expressed either unconditional or conditional support for the Cluff Lake uranium mine.* Half of them expressed support for a moratorium, which shows widespread grassroots support for the official position taken by Indigenous organizations. About 25% were out rightly opposed to the uranium mine and another 25% were neutral. An interesting and ironic finding was that 83% of the northern proponents of uranium mining were from Uranium City, which has since become depopulated due to the shutdown of its uncompetitive mines after the start up of the Cluff Lake mine.

Indigenous people clearly wanted to take the CLBI seriously. First they lobbied for Aboriginal rights to be considered in the Inquiry. Though their official organizations ended up boycotting the CLBI's formal hearings because they would not address Aboriginal rights, a strong grassroots Indigenous voice calling for a moratorium persisted in the local hearings. With clear historical precedent for recommending a moratorium, the CLBI chose instead to recommend revenue sharing with the north — which the NDP government rejected. The temptation for the NDP government to cash in on the nuclear energy boom they hoped for in the wake of the looming oil crisis was too great for them to be distracted by issues of fundamental justice. It was the beginning of a politics of manipulation and denial that was to go from bad to worse.

Notes

1. Jim Harding, "Development, Underdevelopment and Alcohol Disabilities in Northern Saskatchewan," *Alternatives* Autumn, 1978, pp. 30–39.
2. Jim Harding, "Due Process in Saskatchewan's Uranium Inquiries," in Dawn Currie and Brian McLean (eds.), *Rethinking the Administration of Justice*, Fernwood, 1992, pp. 130–49.
3. *Final Report of the Cluff Lake Board of Inquiry*, Regina, SK, 1978, p. 204.
4. R.H Bartlett, "Indian and Native Rights in Uranium Development," *Saskatchewan Law Review* 45: 1, 1981, pp. 14–15.
5. Jim Harding, "A Content Analysis of Attitudes Towards Uranium Mining Expressed in the Local Hearings on the Cluff Lake Board of Inquiry," *Impact Assessment Bulletin* 4: 1–2 (Special Issue based on the North American Conference in Calgary), 1986, pp. 189–209.
6. Meadow Lake and Prince Albert District Chiefs, *Statement*, October 7, 1977.
7. The statements from AMNSIS are from CLBI transcripts, pp. 5740–44.
8. "Scrip" refers to a provisional certificate entitling the holder to a share of property or land.
9. Clem Chartier, "Aboriginal Rights and Land Issues: The Métis Perspective," in M. Bolt and J.A. Long (eds.), *The Quest for Justice: Aboriginal Peoples and Aboriginal Rights*, Toronto, University of Toronto Press, 1985, pp. 57–58.
10. See Jim Harding, "Why Louis Riel University," Simon Fraser Student Society, Special Issue, *The Peak*, July 1968.
11. CLBI transcripts, p. 5746.
12. For an in-depth study of this colonial process, see Dean Neu and Richard Therrien, *Accounting for Genocide*, Fernwood, 2003.
13. The following quotes are from *Native Policy: A Review with Recommendation*, May 27, 1976, pp. 1, 5, 3, 2, 2 and 1–2 respectively.
14. D. Gruending, *Promises to Keep: A Political Biography of Allan Blakeney*, Saskatoon, SK, Western Producer Books, 1990, p. 120.
15. Jim Harding, "A Content Analysis."

chapter 2

GOVERNMENT DISINFORMATION CAMPAIGN

The Saskatchewan NDP government succeeded in keeping uranium mining and nuclear energy from becoming an issue in the 1978 provincial election. The NDP talked in generalities about its resource policy, which it promoted with a mix of nationalistic appeal and a slick media defence of its "publicly owned" Crown corporations. The nuclear issue was effectively sidelined into the low profile public hearings of the Cluff Lake (Bayda) Inquiry (CLBI), and the Saskatchewan NDP was re-elected without having to justify its uranium policy. The NDP's electoral strategy emphasized public ownership of uranium as a means to ensure benefits for the province. Tommy Douglas was even brought to some election campaign rallies to support Allan Blakeney, who talked about Saskatchewan becoming a "have province" as though uranium was just another natural resource that could help build the social democratic promised land. This "golden egg" policy ensured a lot of trade union support and split the political left. At one point the Saskatchewan Waffle was still asking whether nuclear power would be acceptable under socialism.

Covert Action by the Uranium Secretariat

Meanwhile those driving the uranium policy were readying themselves for the coming battle over public opinion. In 1979 the NDP government created the Uranium Secretariat within the Executive Council. Operating until 1982, when the Blakeney government was defeated, it was mandated by the Cabinet's Uranium Committee and provided with a coordinator, Jack McPhee, who was deputy secretary of the Planning Commission of Cabinet. Over its four years it coordinated efforts among the Crown corporations and regulatory agencies, monitored the uranium inquiries, undertook surveys, helped with NDP convention floor strategy and, after 1981, monitored anti-nuclear groups and magazines.

This was a clear demonstration that the Blakeney NDP government's attempt to shelve the nuclear issue had failed. Interest in and concern about nuclear energy in Saskatchewan had steadily grown, sparked by events such as the reactor accident at Three Mile Island on March 28, 1979, and the timeliness of the film *The China Syndrome*. Opinion polls revealed that 63%

of Canadians were opposed to the expansion of nuclear energy. The NDP's own Gallup surveys in the province from 1976 to 1980 were not encouraging, showing that support for uranium mining slipped from 72% to 64%, support for a uranium refinery slipped from 65% to 54%, and, perhaps most notably, support for a nuclear power plant slipped from 63% to 36%. With public support for its long-term nuclear strategy declining and worried about Greenpeace entering the fray, it intensified its activities, raising the profile of the Heritage Fund, holding more briefing sessions to create more "government sales reps," utilizing direct mailings to community leaders and placing pro-uranium stories in the party paper, *The Commonwealth*. It established the Mass Media Coordinating Committee, in which even the regulator, Saskatchewan Environment, was involved and held focus groups with the aid of Atomic Energy of Canada Limited (AECL) and Ontario Hydro to help it refine its propaganda.

The Uranium Secretariat sometimes directly reported to the Premier, as when McPhee tellingly wrote in 1981 that, "The reactor issue is not moveable in this time period without great risk."[1] The Secretariat's activities were meant to remain covert, with Premier Blakeney denying its existence shortly after it was formed. In secret it launched the Energy Teaching Aid program, modelled on Ontario Hydro's, because, "The short term benefit that will be had from this program is that in the next year we should have a number of teachers across the province 'inside'." It also intervened directly in the public controversy through its Speaker's Bureau, what became known as its Truth Squad and a government newsletter, *Uranium Report*.[2] In other words, the NDP government ran a covert pro-nuclear campaign permeating government and civil society that, in retrospect, looks more like an authoritarian than democratic state apparatus.

The political objective of the government's new agency was to try to undercut the increasing appeal of the non-nuclear position among the public — at the cost of an estimated $400,000 to the Saskatchewan taxpayers. The contract for devising this pro-nuclear campaign went to Struthers and Associates, a Regina advertising firm that had served the NDP well. Struthers and Associates had recently been purchased by Russ Eaton and Cam Cooper, men with a long record of working in the NDP bureaucracy. (At the time Cooper was also a columnist with *The Commonwealth*.) This change of ownership in the firm receiving the government contract signalled a tightening relationship between the NDP and its political advertising consultants. The Executive Council of the NDP government had already hired Lowell Monkhouse, former manager of the Struthers firm, to prepare recommendations for the coming pro-nuclear campaign.

Details about this government campaign were carefully guarded and remained top secret until it was launched, much as the decision that Crown corporations would buy into the uranium industry was also kept from public scrutiny until the deals were completed. While it was suspected that some government actions were being coordinated at the time, I could only confirm the extent of this covert action by the Uranium Secretariat after in-depth research on the policy background to uranium mine expansion was completed in 1997.[3]

What You See Isn't What You Get

After January 1980, the Uranium Secretariat created a Speaker's Bureau to promote uranium mining while appearing to be autonomous from government and industry. By late 1981, it had nineteen members, mostly natural scientists from the province's two universities. Interestingly, the speakers included two former commissioners from the CLBI (Angus Groome and Ken McCallum) and pro-uranium academics from the University of Saskatchewan's Uranium Research group, in existence since 1979. Speakers were sent to service clubs, professional associations, rural municipalities and NDP riding associations as though they were detached educators with an expertise on this issue. A major goal was to broaden the audience getting the pro-nuclear message by getting press coverage for these appearances. Uranium Secretariat staff kept a low profile while they monitored these events and later met with speakers to try to improve performances. Travel expenses, communications training, research backing and news summaries were provided to members by the Secretariat, with "most costs covered by the SMDC," the Crown corporation undertaking uranium expansion. The SMDC and the AECL also provided free tours of uranium mines and nuclear power plants.

From August to November 1981, the Uranium Secretariat sent speakers to nearly thirty events, and requests kept rolling in. One of the most active members was Wayne Larson, who was particularly adamant in his criticisms of the United Church and other opponents. He made very questionable claims — that low level radiation was not harmful, that nuclear power is cheaper than other energy sources, that there was no connection between uranium mining and nuclear weapons and that nuclear wastes were being handled effectively. Under the pretence of being independent academics with an educational mission, he and other university employees were *de facto* lobbyists for the nuclear industry.

"The Truth Squad," formed in March 1982, was inspired by the public relations promotions of U.S. nuclear reactor manufacturers and utilities, including tactics such as crashing events organized by anti-nuclear groups. The Truth Squad had thirteen members, five from the Speakers Bureau, five from the SMDC and three from the Secretariat.[4] The main targets were church groups, particularly the Inter-Church Uranium Committee and the

United Church, after it censured the government's uranium policies at its June 1981 conference and launched a series of workshops on the controversy in February 1982. Truth Squad members went to these workshops to disrupt the agenda and challenge the speakers. Even the right of the church to take a stand on this issue was challenged — an unusual state-initiated affront to the separation of church and state. The Secretariat helped with this campaign by mailing over 300 copies of a seemingly pro-nuclear Methodist Church of Britain document to church congregations throughout the province.[5] These interventions in church events were often upsetting to local congregations, but they succeeded in hijacking publicity, which was one of the Truth Squad's objectives.

Government's *Uranium Report*

The only thing we could get a handle on at the time was *Uranium Report*, a newsletter distributed widely in an effort to consolidate party and public support for uranium expansion. But the danger in putting pen to paper is that positions can then be publicly scrutinized. Because of my background in environmental studies, I found the disinformation in the *Uranium Report* particularly offensive, so I began to monitor the monitors. It soon became clear that the uranium proponents felt vulnerable on two big questions: the environmental health hazards of uranium mining and the potential of alternative renewable sources of energy. The NDP government's uranium newsletter consistently took positions that ignored or misrepresented the developing research in both these fields. Looking back to the propaganda campaign of the late seventies from the present position of a Lorne Calvert-headed NDP government trying to "green" the image of the party, we can see how far from progressive democratic practices the NDP government and bureaucracy of the time were willing to go in defence of what is now clearly indefensible.

Uranium Report, issued from the offices of Jack Messer and Elwood Cowley, the ministers responsible for mineral resources and the uranium Crown corporations, was subtitled, "A newsletter dealing with your government's activities in the uranium industry." The newsletter gave a reassuring picture of the environmental regulatory process in Saskatchewan. Volume one, number two stated, "Large developments must be authorized by the Minister of Environment before construction can begin — this is true whether the project is a Crown corporation, private company..." — an assertion that had been publicly contradicted by none other than the government's own attorney general and future Saskatchewan premier, Roy Romanow. The premier-to-be acknowledged that the Minister of Environment hadn't approved the building of canals and the drainage of Key Lake by the joint government-industry uranium mining venture,[6] and leaked government memos printed in the October 1979 issue of *Briarpatch* magazine confirmed that the joint venture taking shape during explorations for uranium was "technically in trespass."

The government newsletter effectively misguided the public about health and safety, omitting the mention of the 300,000 tonnes of radioactive wastes produced for every year of reactor operation, wastes that will remain toxic for thousands of years. By only comparing the annual fuel needs of a 1,000-megawatt (MW) nuclear reactor (e.g., 30 tonnes uranium) to the much larger tonnage of fuel required by other forms of energy (e.g., oil, coal), the newsletter also falsely asserted that nuclear energy is environmentally benign. This was the beginning of the "biggest lie" yet in the nuclear controversy. The newsletter also failed to mention that most of the radioactive waste would remain in the uranium tailings to accumulate, in this case, at the new Cluff Lake and Key Lake mines. This was no laughing matter, as an estimated 80 millions tons of such tailings already accumulated at Elliot Lake, Ontario, have contaminated the nearby Serpent River system. Other radioactive waste would come in the form of the toxic spent fuel and irradiated coolant water, which at the time were accumulating at reactor sites throughout the world. In 1979 it was widely acknowledged that the industry had yet to find a safe method for storing these wastes; the same technological "leap of faith" that was first made when nuclear weapons and nuclear energy were developed after World War II was being made in Saskatchewan under the NDP government.

The government newsletter also claimed that radioactivity from uranium mining was fairly innocuous compared to other sources. It stated, "The total amount of radon coming from all mines is minute when compared to natural radon emissions from Canada's land surface." This sort of comparison was common among nuclear proponents. Medical radiation researchers close to the nuclear industry often emphasized that radiation from radon (called alpha) can be stopped by human skin. However, they failed to mention that once the natural uranium ore is mined, crushed and left as uranium and tailings, this alpha radiation can more easily enter the human lungs as dust. When alpha radiation is inhaled it is more likely to cause lung cancer than other forms of radiation (e.g., beta, gamma).

Elsewhere the government's *Uranium Report* noted that 60% of Canada's hard rock drilling — mainly for uranium — was occurring in Saskatchewan. Yet there was no mention of the substantial increases in radioactivity in the air, water and food chain that would occur from these massive uranium ventures. Number 3 of the newsletter also stated as an "energy fact" "that "natural gas contributes 100 times more radioactivity to the public than does the nuclear fuel cycle." No scientific sources were provided for this assertion; the public was being asked to take these assertions on faith from government departments that were strongly promoting the nuclear industry. But this government "fact" was contradicted by none other than the U.S. Environmental Protection Agency, which in February 1978 stated that from all available data, the largest radiation dose for individuals received from all sources came from radioactive material added to the environment. This was "mainly as a result of inhalation of radon daughter products from uranium mine tailings."[7] Even when radiation to the general population, rather than to individuals, was calculated by the U.S. Environmental Protection Agency, it was found that after medical x-rays (at the time a serious problem in its own right) the greatest source of radiation was from radioactive material added to the environment — mainly from uranium mining and milling, radon in water and in construction materials, and from natural gas and geothermal energy.

From the perspective of social and preventive health, the real threat to the public comes from the accumulation of radioactivity in the environment and its interaction with other accumulating industrial toxins. Public health specialists in the U.S., such as John W. Gofman, MD, have pointed out that while some radiation comes from natural gas and coal, this is no excuse for creating significantly higher, longer-lived levels of radioactivity from uranium mining and nuclear energy.

Exxon and NDP Oppose Effective Conservation

Skirting the public health issue, the government's nuclear push used the argument that all ventures entail risks and risks have to be taken because of the impending energy crisis. But a close reading of the newsletter's Number 3 discussion of solar hot water heating undercuts this argument. In an attempt to leave the impression that solar energy cannot substantially decrease the demand for a non-renewable resource like uranium, the newsletter stated, "if all households in North America were converted to solar water heating… it would save only 4 percent of our energy requirements." But the newsletter never mentions that, besides building bombs, uranium can be used *only* to generate electricity, and that nuclear energy at the time *only* accounted for 13% of total U.S. and 8% of total Canadian electrical energy. Furthermore, electricity is only one component — about one-third — of total energy use.

These omissions veil the reality behind the assertion that solar water heating can only provide 4% of North American energy — when the reality is that *nuclear energy itself was only able to provide less than 4% of total North American energy requirements*. Since this was all for electricity, an inefficient but major source of home water heating, solar water heating had — and still has today — the potential for substantially decreasing, and even replacing, much of the demand for uranium and nuclear energy.

Since some of the government's "energy facts" disproved their own case, how was their pro-uranium campaign to be understood? The NDP government had clearly chosen to support the view of the multinationals that new energy sources must come from "their" non-renewable supplies. The government newsletter notably failed to mention the "energy fact" that many of the same large multinationals (e.g., Exxon, Gulf) now controlled large amounts of all non-renewable energy, including oil, coal, gas and uranium. These corporations — and, it seems, the NDP government — wanted the public to believe that their only choice was between radiation from nuclear power and acid rain from coal-fired generating plants. Energy conglomerates and the government alike stood to profit, whichever form of non-renewable energy was chosen. Presenting such a false dichotomy is the same tactic being employed today, with government and industry making it appear as though our choices are between global warming and nuclear energy.

Even in the seventies, energy future experts were suggesting that the real choices were between conservation and renewable energy systems developed in the public interest (i.e., not owned and controlled by multinationals) or hazardous non-renewable resources. Yet there was no mention in *Uranium Report* of this most basic ecological choice. Instead, we find the statement in number 2 that "conservation does not solve the problem, it merely buys time. A 25 percent saving would only serve to increase existing reserves by 25 percent."

The false pessimistic choice set up by the government was seriously contradicted by reputable studies. A 1979 study conducted at the Harvard Business School[8] calculated that solar energy could provide 20% of U.S. energy needs by the year 2000 and that a further 30–40% of energy usage could be cut, with no decline in the standard of living, through concerted conservation programs. But the Saskatchewan NDP government, having become fully integrated into the multinational nuclear industry, had already spent $600 million of the taxpayer's money on uranium mining. No wonder it couldn't fathom the emerging "green" position stressing a safe and sane energy future of conservation and renewable energy in a self-sufficient society. Informed sources close to the government have confirmed that the decision to expand uranium mining was quickly taken prior to the public inquiry, in the hope of reaping a bonanza from Saskatchewan's high grade ore, taking advantage of the inflated world price set by the uranium cartel. Only later, as international opposition to nuclear energy mounted and alternative energy sources became realistic, was the NDP government forced to develop public justifications for their fundamentally opportunistic decision.

The government's *Uranium Report* may have partly met its political objectives within the party membership. It was sent to most NDP constituency executives prior to the 1979 convention and probably played a role in maintaining support among most delegates for continuing the government's uranium policy. The government had used propaganda techniques — taking information out of context and distorting or omitting the facts — in its uranium newsletter, but as independent sources of information slowly became more available in the province, resistance to uranium mining and refining continued to grow.

Notes

1. Jack McPhee to J.E. Sinclair, Deputy Minister to the Premier, May 28, 1980. Jim Harding and Dave Gullickson, *Provincial Policy Chronology of Uranium Mining in Saskatchewan*, In The Public Interest, Research Report No. 3, Prairie Justice Research, University of Regina, 1997, p. 110.
2. The SMDC issued its own newsletter but collaborated with the Uranium Secretariat on various initiatives, such as the Speaker's Bureau.
3. Jim Harding and Dave Gullickson, *Provincial Policy Chronology*.
4. These included Agnes Groome of the CLBI, Wayne Larson of the University of Regina, Rita Muirwald and Dale Schmeichel of the SMDC, and Lowell Monkhouse and Jack McPhee of the Secretariat.
5. *Shaping Tomorrow*, Home Mission Division, Methodist Church of Britain, 1981. While this document did treat nuclear energy and nuclear weapons as separate issues, more recent activities of the Methodist Church suggest it does not now support either.
6. See November 3, 1979, Saskatoon *Star Phoenix* and November 7, 1979, Regina *Leader Post*.
7. Environmental Protection Agency, *Natural Radio-Activity Contamination Problems*, Feb. 1978, p. 4.
8. R. Stobaugh and D. Yergin (eds.), *Energy Future: Report of the Energy Project at the Harvard Business School*, Random House, 1979.

chapter 3

CHALLENGING NUCLEAR SCIENTISM

With the expansion of uranium mining seemingly secure after the NDP returned to power in 1978, the Saskatchewan NDP government turned its attention to getting a uranium refinery in the province. The purpose was to create "value added" activities in the resource and energy sector as a means to diversify the economy. As a potential "have not" province, with shrinking federal transfer payments, there seemed no limit to what it was willing do to attract the nuclear industry. The province later moved to get a Candu nuclear reactor and flirted with establishing a nuclear waste industry.

The federal Crown corporation, Eldorado Nuclear, picked Warman, a Mennonite town near Saskatoon, as one possible site for a uranium refinery that would be closer than its existing one at Port Hope, Ontario, to the mammoth uranium mines in Saskatchewan's north. The federal hearings on the Warman proposal were held in January 1980; they became a lightning rod for the frustrated and evermore cynical non-nuclear movement in Saskatchewan. The resulting coalition against the uranium refinery, which brought together non-nuclear and Mennonite groups, was to become the most broad-based and effective environmental movement seen to date in Saskatchewan. The wide scope of opposition was expressed in the book, *Why People Say No!*, published by the Regina Group for a Non-Nuclear Society in 1980. In the end, the panel overseeing the hearings turned down the proposed refinery, primarily for social impact reasons. Warman was to be the first victory in the emerging non-nuclear movement in the province.

Through participating in the public hearing process that stopped the Warman refinery, the non-nuclear public experienced a huge learning curve about the makeup of the whole nuclear fuel system, which has stayed with it to the present. The refinery opponents were adamant that the full-scale implications of uranium refining for human health must be taken into account if the conclusions were to be credible. Industry and government tried to separate the assessment of the proposed refinery from the controversy over uranium mining, and nuclear reactors and weapons, which further infuriated opponents. Because of my background in teaching environmental health and in health research, the Warman and District Concerned Citizens asked me to present the public health concerns to the federal hearings.[1]

In an era when technological science has become the handmaiden of industry and government, the environmental dimension of public health often goes unexamined. It is common practice for technical evaluations of idealized models to examine only fragments of the nuclear fuel system, then to perpetuate the findings as scientific truth. This is not science, but "scientism," where experts employed by the nuclear industry keep the discussion of health and safety inside this pseudo-scientific format. The political and bureaucratic, rather than ecological, terms of reference for government inquiries usually work to the industry's advantage.[2] The total accumulating environmental and human health hazards of mining, milling, transportation, refining, enrichment, reactors, weapons and nuclear waste storage are ignored, and the nuclear industry expands, bit by bit, with only piecemeal scrutiny.

The use of scientism in shaping public opinion was exemplified in Eldorado Nuclear's obfuscation of the health hazards of nuclear energy as presented in its Environmental Impact Statement (EIS) for the proposed refinery. It is true, as the EIS stated, that yellowcake and refined uranium hexafluoride contain lower levels of radioactive material than exist at the earlier mining, or later reactor, stages of the nuclear fuel system. But it was not truthful for Eldorado to restrict the discussion of the health hazards to a hypothetical and technical consideration of radioactivity and other chemical pollutants coming from the proposed single plant only. This, at best, was nothing but an academic exercise.

The yellowcake transported for refining can't be separated from the earlier mining and milling activities in northern Saskatchewan. Prior to these hearings it was estimated that for each year of production of uranium fuel for a 1,000 MW nuclear reactor there will be 300,000 tonnes of long-lived radioactive wastes created, most of these as mine tailings.[3] Furthermore, the uranium hexafluoride shipped from any refinery to enrichment plants abroad will be used at some point in reactors and/or weapons' production. It is the stuff out of which comes the concentrated carcinogen plutonium, with a half-life of 24,400 years, which accumulates across the globe.

Many officials employed at different stages in this fuel system believe in their fragmented models. Institutional roles and authority, peer and career pressures, and class interests can, and do, narrow officials' frames of reference. We saw during World War II, for example, the devastating effect that narrow terms of reference had on scientists who, with few exceptions, submitted to the particular nationalisms of their geographic location and upbringing. As a technology-driven science becomes more dependent upon corporations with their profit motives and governments with their socio-political control preoccupations, the appropriation of science becomes even more pervasive. And with that comes a new form of propaganda — scientistic propaganda. Nowhere is this more apparent than in the nuclear industry, which has its origins in the military, remains dependent on massive government subsidies and is surrounded by a cloud of official secrecy.

Warman Refinery: Testing the Environmental Review Process

The opponents to the proposed refinery had already found out that public inquiries, whose only authority was to report to governments already deeply committed to nuclear energy, were hardly a method of ensuring the public interest. Furthermore, the scientific quality of deliberations was highly questionable. The Cluff Lake (Bayda) Report, for example, did not consider recent research on the far greater than expected health hazards of "lower" levels of radiation common in the nuclear industry. According to the chair of the 1980 B.C. Uranium Inquiry, which led to a seven-year moratorium of uranium mining in that province, the Bayda Inquiry did not even calculate the health hazards of alpha radiation to miners.[4]

Eldorado's EIS for the proposed uranium refinery continued to accept standards, such as 5 rem/year to a worker, that had not fundamentally changed since 1958. These standards had been set before intensive epidemiological research on the cancerous and other effects of exposure within these limits were even possible. Meanwhile, many radiation researchers were advocating the reduction of radiation standards by a factor of five to ten.[5] Pressure for such a reduction continues to build with more recent research. Furthermore, such standards assume normal operating conditions — but nuclear contamination accidents never occur under assumed conditions. With a cumbersome technology like nuclear energy the chances of technological malfunctions and human errors are increased. If the legal radiation standards had been lowered to become preventive health standards, it would have made the nuclear industry even more uneconomical at a time when it was already having serious economic problems.[6] Demand for Eldorado's refined uranium would decline. It was open to question whether the mining operations providing yellowcake could continue if public health radiation standards were created and enforced.

It was clearly in Eldorado's interest to maintain the status quo in the radiation standards of the time, standards which several eminent radiation researchers claimed allowed a doubling dose for leukemia in just one year of workplace exposure.[7] Even if Eldorado had doubted the converging conclusions of independent studies on lower levels of ionizing radiation, the only responsible, rational and moral response was to postpone any expansion of the nuclear fuel system until the scientific dispute was sorted out. This should have led to an outright moratorium on the proposed Warman refinery. Otherwise, the government totally contravened the tenets of preventive public health.

In the 1970s Dr. Alice Stewart of England, a world-renowned public health researcher, discovered that children exposed *in uteri* to supposedly small typical x-ray radiation during the first trimester of pregnancy were ten to fifteen times more likely to contract leukemia in their first ten years of life, than those who were not so exposed. She had also worked on the Hanford study showing higher than expected rates of cancer among nuclear plant workers.[8] Her concluding remarks to the British inquiry into the Windscale nuclear reprocessing plant were:

> If our conclusions are not completely accepted, then it should be clear that the evidence relating to the safety of the public... is disputed among experts, and consequently... no decisions should be made until the issues have been satisfactorily resolved.

It is notable that nearly three decades later the allowable exposure to radiation for Saskatchewan uranium workers remains 5 rems (or 50 millisieverts — mSv) per year, the same level set in 1958. This level is set by the *Radiation and Safety Act* of 1985, updated in 2005, which takes its limits from the International Commission on Radiological Protection (ICRP). In 1990 the ICRP lowered its five-year exposure for occupational workers to 10 rems (or 100 mSv), and Saskatchewan followed suit. However, in 2005 the Biological Effects of Ionizing Radiation Committee (BEIR VII) concluded there was no safe level of radiation exposure, and the Canadian Nuclear Safety Commission now admits there is a "small risk of childhood cancer above 10 mSv" a year.[9] It is noteworthy that a person in Saskatchewan who is not an occupational worker is only allowed exposure to 1 mSv in a calendar year.

Public Skepticism of the Nuclear Industry

Traditionally, the Canadian public has viewed government environmental inquiries as not particularly geared to serve their interests, and this is certainly true in the arena of nuclear energy. The May 1979 Gallup Poll showed a full 63% of Canadians opposed any further expansion of nuclear energy, and support for nuclear energy had dropped 18% over the previous four years, while opposition to it had increased by 29 percent. As the voice of the non-nuclear movement became louder, undecided Canadians were clearly going non-nuclear.

Public skepticism of the nuclear industry had grown, in part, because of the shoddiness of the regulatory system and its biased pro-nuclear pronouncements on health and safety. A vivid case was the 1975 Rasmussen Report on reactor safety commissioned by the U.S. Nuclear Regulatory Commission. This study, which calculated that the chance of a core melt-down was 1 in 100,000 per year of reactor operation, had to be recalled by its originators even before the Three Mile Island partial core melt-down in 1979 totally disproved its ideal model. Yet the CLBI used the Rasmussen Report in 1977 to "ethically" justify the export of Saskatchewan uranium. The report had already been discredited after the Union of Concerned Scientists obtained secret background papers, under the U.S. *Freedom of Information Act*, which showed that the design and results had been deliberately fixed to be supportive of the nuclear industry. These documents, published in the mid-February 1978 issue of *Nuclear Blowdown*, also revealed that the study's authors were nuclear engineers directly tied to the nuclear industry and that reactor safety issues that would "not support (its) predetermined conclusions" were totally ignored.

If Canada's *Access to Information Act* (1990) had been in effect at the time, the public skepticism of Eldorado probably would have increased. With access to only its carefully edited publications on the proposed Warman refinery, even without the secret background information, there were clear reasons to fundamentally question Eldorado's concern for public health. Page one of Eldorado's widely circulated pamphlet, "Uranium and Electricity," stated, "Since no combustion is involved, nuclear fuel offers an environmentally clean and efficient method of power production in contrast with burning coal or oil to generate electricity." In spite of widespread criticism, Eldorado stuck to its guns, repeating the statement in its EIS for the proposed refinery, not even bothering to qualify the "clean" nature of nuclear energy in comparison to coal and oil. Whether through ignorance or through propaganda, the statement completely distorted common knowledge about the health hazards of radioactivity. It also evaded the fundamental challenge made by many scientists about the inefficiency of nuclear energy compared to renewable energy and conservation.

Nuclear energy may operate outside our human senses and appear clean in comparison to fossil fuel combustion, but that makes it more, not less, hazardous to public health. Because ionizing radiation operates outside our biological senses and because there is a long latency period (sometimes decades) before cancers become clinically diagnosable, it is a most insidious form of environmental pollutant. Eldorado officials could have easily found this out. Evidence was presented to a 1978 U.S. congressional seminar indicating that while cancer rates were dropping in some visibly polluted cities, they continued to rise in areas that are near visibly "clean" nuclear plants.[10] It has also been suggested that negative synergistic health effects may result from increases in radioactivity levels, resulting in the weakening of our immune systems, making us more vulnerable to other toxins or infectious agents.[11]

Presenting a nuclear-based economy as "clean" is the ultimate form of scientific forgery. It is reminiscent of the linguistic tactics described by Orwell in *1984*. It is encouraging to contrast this kind of irresponsibility with the activities of medical professionals concerned about the public health hazards of nuclear energy. Just prior to the Warman inquiry, in May 1979, Doctors R.F. Woolard and E.R. Young of the B.C. Medical Association Environmental Health Committee published an excellent annotated bibliography in the BCMA journal on the health hazards of the nuclear fuel system. Unfortunately, the Saskatchewan Medical Association did not heed the call of their B.C. colleagues and invited only pro-nuclear spokespeople to their 1979 convention — seemingly to hear reassurances from the industry and to savour the benefits the "uranium boom" would bring to those who were heavily involved in real estate in Saskatchewan.

Yellowcake Tailings and Public Health

The struggle over the proposed uranium refinery convincingly demonstrated that the health hazards of uranium needed to be studied in their total real-world context. The hazards from tailings left after the production of yellowcake were thought in 1979 to present the greatest overall threat to public health in the whole nuclear fuel system. The renowned researcher John Gofman suggested that past estimates of the releases of radon gas from the nuclear fuel system may be underestimated by a factor of 100,000.[12] Recalculations based on the updated figures led to an estimate of 100 to 400 deaths for every day that the nuclear industry was allowed to continue to operate in the U.S. Such underestimations were used in the Cluff Lake Inquiry and continue to be used in the Saskatchewan government pro-nuclear public relations campaign.

By 1979 there were already enough tailings in the U.S. to cover a four-lane highway coast-to-coast to a depth of one foot. When uranium mining began, the public was predictably assured that the radioactive tailings could simply be placed back in the mines, but these technological fixes lacked ecological common sense and foresight. Since the milling process increases the volume of material, it proved impossible to get them back into the mine holes. In any case, most of the new Saskatchewan mines are of the open-pit variety with potentially greater exposure of radionuclides to the air, water and food chain.

Proponents of the nuclear industry continue to tell us that since radiation is natural there is no reason to be concerned about the effects of radioactivity from the nuclear fuel system. The Saskatchewan NDP government *Uranium Report* claimed that we have more to fear from natural uranium ore and even natural gas than uranium mining. Meanwhile, the U.S. Environmental Protection Agency (EPA) had in a February 1978 report, *Natural Radioactivity Contamination Problems*, noted that "public health problems" with radioactive material were usually the result of "the 30 or more radio-nuclides in the uranium and thorium decay series because of their relative abundance and toxicity." The EPA specifically mentioned the risks to uranium workers and, in a May 1976 report, *Radiological Quality of the Environment*, highlighted "increased population exposure (from) radon problems in Western States due to the use of radioactive tailings." We can go to any region of the world where uranium mining has occurred and find clear evidence that the nuclear industry threatens public health. In Canada, the Ontario Serpent River basin is being contaminated by seepage from 130 million tonnes of tailings left from fourteen decommissioned uranium mines near Elliot Lake. In 1993 a review panel agreed these constitute a long-term environmental hazard; however, disputes continue over how to undertake restorative measures. Dr. Charles Kerr, a professor of social and preventive health from Australia, informed the B.C. Uranium Inquiry in 1980 that there had been a dam break and spill of massive tailings in his country in the 1950s.[13] At the time details were still not available to public health researchers because of the protection of the nuclear industry there, as here, by official secrets legislation. The B.C. Royal Commission on Uranium Mining heard evidence that the overall mining industry did not have a good record of preventing tailings dam spills and that in British Columbia, where there are no uranium mines, there was an average of one *recorded* tailings dam spill per year since the 1970s.

Tailings Spill in New Mexico

The mid-February 1978 issue of *Nuclear Blowdown* reported that just a year prior to the Warman hearings, on July 16, 1979, the tailings dam at the Church Rock uranium mine in New Mexico broke, releasing 100 million gallons of radioactive water and 1,100 tons of radioactive tailings in a flash flood that travelled sixty miles down the Rio Puerco River before being absorbed into the desert. As in Serpent River, the immediate victims were Indigenous people, in this case Navajo, whose farming was threatened because their livestock drank from the river.

One of the contaminants in the spill, thorium, is highly carcinogenic and is permanently absorbed into the body. It has a half-life of 76,000 years and is toxic for ten to twenty times longer. Fearing that the health effects on the Navajo children, who regularly played by the river, may not show for decades, a local pediatrician was publicly critical about the feeble attempts to clean up the spill. He noted that workers, without safety masks and using nothing but hand shovels, had recovered less than 1% of the spillage.

The company claimed that the tailing pond "was designed according to the best engineering practices" and blamed a "unique subsurface bedrock configuration" for the spill. This is reminiscent of the way Reed Paper, of Dryden, Ontario, tried in the 1970s to deflect its responsibility for regional pollution by blaming riverbed plant-life for transforming the spills of elemental mercury from their paper plant into dangerous methyl mercury.

The strategy of the U.S. mining company United Nuclear was to allow "natural" dilution of the radioactivity until it was within the legal, permissible limits. Sulphate, considered an early warning system of groundwater contamination, had already increased in local wells. The Church Rock spill was within twenty miles of a national park that attracted nearly a million people a year. The regional conflict over uranium mining continues to this day. In 2006 Navajo groups were organizing to oppose Hydro Resources' plan, with uranium at $50 a pound, to use *in situ* methods to mine uranium near Church Rock.

New Mexico, like northern Ontario and northern Saskatchewan, had been selected as an international sacrificial area of the nuclear industry. Seventy million tons of uranium tailings had already accumulated in New Mexico, where, according to evidence given to the B.C. Uranium Inquiry, there were "three uranium tailing piles within 10 miles of each other, each one over 20 million tons... leaking like sieves." The evidence also showed a twelve-fold increase in alpha radiation and a ten-fold increase in uranium concentrations in local groundwater.

Other End of the Fuel System

During the controversy over the proposed uranium refinery at Warman, the non-nuclear movement convincingly showed that refinery operations depend upon uranium mining and tailings, which can hardly be considered "clean." How clean did they find nuclear energy to be at the other end of its fuel system? In May 1979 at the Sixth International Congress of Radiation Research in Tokyo, Dr. Carl Johnson, a public health official from Colorado State, reported his findings about cancer rates among people living downwind from the Rocky Flats nuclear plant. The cancer incidence in the downwind population was significantly higher than in nearby unexposed areas. In a three-year period, rates of lung, leukemia, lymphoma and myeloma cancers were 40% higher for men in the downwind area.

The end product of the nuclear fuel system, plutonium, is handled at Rocky Flats, U.S., where nuclear engineers had major problems living up to their fail-safe utopia regarding waste containment. Making a technological "leap of faith" years ago about an imminent solution to nuclear waste disposal (the same "leap of faith" made by Eldorado and Saskatchewan Premier Blakeney in the 1970s), Rocky Flats ended up with 5,500 barrels of plutonium stored in oil by 1969. That year a major fire occurred in the plant, and the regulatory body, the Atomic Energy Commission (AEC), was called in to determine if there had been any loss of plutonium. It finally reported that a milligram had escaped, but a skeptical scientist decided to take his own independent measurements, finding traces of plutonium miles from the plant and concluding that one pound, not 1 mg, had actually escaped. This was later confirmed by the AEC's health and safety staff, which showed that the official regulatory agency underestimated plutonium traces by a factor of 200,000. To further tarnish the credibility of the nuclear regulatory system, it turned out that the plutonium releases had not come from the fire at all. Rather, 25% of the barrels storing plutonium had rusted out. Half a pound of plutonium (enough to give 250 million people cancers if evenly distributed among humans) ended up in the greater environment. Some was taken by the high winds nearly as far as Denver.[14]

When this was raised in the refinery hearings, Eldorado Nuclear protested that their proposed refinery at Warman wouldn't store plutonium, arguing that what happened or happens at Rocky Flats was irrelevant. But is there anyone who knows with confidence where all the spent fuel from the uranium that has come out of the Eldorado mine at Uranium City since the 1950s has actually ended up? Much of it certainly found its way into the U.S. nuclear weapons stockpiles, and some of that cancer-causing plutonium in Colorado probably originated from uranium fuel from Saskatchewan.

Can We Learn Anything from Nuremberg?

The pacifist Mennonite community of Warman was particularly concerned about nuclear weapons. Though there has been considerable hair-splitting from the nuclear industry about the exact mechanisms that connect nuclear energy and nuclear arms, the overall historical, economic and technological connection is indisputable.[15] Since nuclear war is the ultimate threat to public health and would constitute genocide against the human race, it is essential for people working in the nuclear industry to consider this possible eventuality. Though the nuclear industry and its potential complicity in the ultimate crime against humanity differs in many ways from the war crimes of Nazism, we can turn to the Nuremberg Tribunal for guidance in understanding personal responsibility in collective crime. According to the Nuremberg Principles,

> Murder, extermination… or other acts done against any civilian population [constitute a crime against humanity]. Crimes against international law are committed by men, not by abstract entities.… [A superior order] does not relieve a person from responsibility under international law, provided a moral choice was in fact open to him.… Individuals have international duties which transcend the national obligations imposed by the individual state.[16]

Extrapolating from the Tribunal's logic, we can say that it is not morally possible to dissociate one's activities at one point in the nuclear fuel system from the consequences at another point. The industry obscures these problems by calling it a "nuclear fuel cycle," but there is no cycle, as shown by the buildup of nuclear wastes. It was exactly this sort of amoral dissociation that was fundamental to Eldorado's EIS on the proposed Warman refinery.

It is clear that bureaucratic, ideological and scientistic rationalizations for the expansion of uranium refining, in view of these principles and the growing worldwide threat to public health posed by nuclear energy and weapons, are at least unjustified if not morally reprehensible. It looks like the people of Saskatchewan will have to relearn these lessons twenty-five years after the failure of the industry to get a refinery at Warman, as the government of recent nuclear-convert, NDP Premier Lorne Calvert, with the full support of opposition leader Brad Wall, goes shopping among the nuclear powers for a new proposal for a uranium refinery for the province.

Notes

1. This chapter is largely based on my brief to the 1980 refinery hearings, also published as "The Public Health Hazards of the Nuclear Industry," *Alternatives*, Winter 1981.
2. J. Harding, "Anti-Nuke Participation Serves Government Propaganda," *Briarpatch*, Vol. 9, No. 3, April 1980, pp. 19–21.
3. J. Gofman, in *Shutdown: Nuclear Power on Trial*, Summertown, Tenn., Book Publishing Co., 1979, pp. 30–31. The amounts will be lower with the higher-grade ore in Northern Saskatchewan, though the tailings will be even more toxic.
4. D. Bates, *Occupation & Environmental Health Considerations*, Nuclear Policy Conference, Carleton University, November 1978.
5. J. Rotblat, "The Risks for Radiation Workers," *The Bulletin of Atomic Scientists*, September 1978, pp. 41–46; R. Bertell, "Measurable Health Effects of Diagnostic X-Ray Exposure," *Testimony before the Sub-Committee on Health & the Environment*, U.S. House of Representatives, July 11, 1978.
6. K.Z. Morgan, "Cancer and Low Level Ionizing Radiation," *Bulletin of Atomic Scientists*, September 1978, pp. 30–40.
7. Doubling dose estimates increases in disease frequency per unit dose of radiation in terms of the baseline frequency of the disease class. It is complex, contentious and regularly being reviewed and updated. See J. Bross, in *Radiation Standards and Public Health, Proceedings of a Second Congressional Seminar on Low-Level Ionizing Radiation*, February 10, 1978.
8. T.F. Mancuso et al., "Radiation Exposures of Hanford Workers Dying From Cancer and Other Causes," *Health Physics*, Pergamon Press 1977, Vol. 33 (November), pp. 369–385. G.W. Ikneal et al., "Reanalysis of Data Relating to the Hanford Study of the Cancer Risks of Radiation Workers," presented at the International Atomic Energy Meeting, Vienna, Austria, March 13–17, 1978. (For a non-technical summary of this research see *Rolling Stone*, March 23, 1978.)
9. <www.hc.sc.gc.ca> accessed Oct. 2006.
10. E.J. Sternglass, in *Radiation Standards and Public Health*, pp. 80–85, 174–82. Also see the interview of Dr. Sternglass by Thomas Paivlich in *Harrowsmith*, No. 28, on the effects of the Three Mile Island partial core meltdown on infant mortality.
11. R. Bertell, "X-Ray Exposure and Premature Aging," *Journal of Surgical Oncology* Vol. 9, 1977, pp. 379–91.
12. J. Gofman, *Shutdown*, p. 7.
13. *B.C. Uranium Inquiry Digest*, No. 6, pp. 13–14.
14. J. Gofman, *Shutdown*, pp. 39–40.
15. This evidence was reviewed in Bill Harding's *Nukenomics: The Political Economy of the Nuclear Industry*, Regina Group for a Non-Nuclear Society, 1979, pp. 29–31.
16. See the Nuremberg Principles at <http://en.wikipedia.org/wiki/Nuremberg_Principles> accessed June 2007.

chapter 4

URANIUM BLOWBACK

After the blatant pro-nuclear bias of the Cluff Lake Board of Inquiry (CLBI) in 1977, the burgeoning environmental and non-nuclear groups in Saskatchewan became even more skeptical of the provincial uranium inquiry processes. Provincial inquiries, unlike the federal hearings on the proposed and overruled uranium refinery near Warman, in which these groups fully participated, were not seen to be at "arm's length." Accordingly, non-nuclear groups decided to boycott the second provincial uranium inquiry, the Key Lake Board of Inquiry (KLBI), appointed in December 1979 to look into the proposed Key Lake mine. It was vital to challenge this mine, which was to be the largest operating mine anywhere in the world. However, as the Key Lake mine was already under development, the results of this inquiry were considered even more predetermined than those of the CLBI. Rather than tie up scarce voluntary resources with the hearings, the non-government groups concentrated on mobilizing more public support for non-nuclear alternatives. The Regina Group for a Non-Nuclear Society (RGNNS) initiated legal action to get an injunction against ongoing lake drainage prior to the inquiry, but their primary focus was to build more links between northern, Indigenous and southern opponents. Activities included a direct action caravan from the south to the north and educational gatherings at the northern village of Pinehouse, on the road to Key Lake. The RGNNS had also wanted to publish a full critique of the Key Lake project and inquiry to follow its book on the uranium refinery at Warman, but this never materialized.

Key Lake Inquiry: Always Looking over Its Shoulder

The KLBI submitted its final report to the Minister of the Environment on January 12, 1981. The Inquiry concluded "that the measures proposed by the Key Lake Mining Corporation (KLMC) are adequate to protect environmental quality… (and) safeguard occupational health and safety."[1] The conclusion — drawn by a government-appointed Inquiry examining a uranium mine that was 50% owned by a Crown corporation of that same government — surprised nobody. But the KLBI had even more credibility problems. For three years before the Inquiry's appointment, the KLMC had been working at the open pit mine site in the north, "dewatering" the Key Lake chain in preparation for mining the high-grade uranium ore that lay beneath it.

Unknown to the public — until government memos were leaked in the summer of 1979 — the KLMC had been doing this work without legal approval from the Water Resources Branch of the Department of Environment. One of the memos, published in the October 1979 *Briarpatch*, stated: "If the concerned public or anti nuclear lobby realizes this... the government would find itself in a rather bad light." As it turned out, the KLBI was appointed while these revelations were prominent in the public's mind. The terms of reference, which stated *it could only recommend how, not whether* the Key Lake mine should proceed, deepened its credibility problems. After studying these terms of reference and in light of the draining of the lakes, the non-nuclear and environmental groups in the province decided to boycott the Key Lake hearings, which created even further problems for the Inquiry's deteriorating public image.

The chair of the Inquiry, Bob Mitchell, had to face the issue of his Inquiry's credibility right off the bat. First, he publicly called for the non-nuclear groups to reconsider and participate, but had no success. One group, the La Ronge Concerned Citizens, attended the preliminary hearings but joined the boycott on March 4, 1980, because the Inquiry overruled its request to have transcripts translated into Cree and Chippewyan, the main languages of the north.

In a further attempt to save its public image, the Inquiry asked the KLMC "to undertake no further work until our Inquiry was completed." This was tantamount to an admission that the lake drainage had seriously undercut the legitimacy of the Inquiry and its assessment procedures. However, the voluntary freeze on the KLMC work at the mine only lasted from February 2 to July 16, 1980, with employees returning to the mine a full six months before the Inquiry was completed, in January 1981. This strongly suggests that the Inquiry's motives were not to stop work at the site while an assessment was taking place but to regain political credibility. The report stated, "Our main purpose for withdrawing our objections at that time was to obtain further data...." A similar circular argument (that the fifteen lakes had to be de-watered to assess the environmental impact of de-watering) was used to justify all preceding work at the mine site.

The Inquiry's final report made several attempts to explain away all of these contradictions. "The Credibility of the Inquiry Process" was claimed to be self-evident because, in the words of the Inquiry, "We knew that we were honest people who were committed to discharging our mandate thoroughly and objectively." It added, almost in desperation, "but how could we demonstrate this to a doubting public?" The personal honesty of Inquiry members was never the issue, although each member always had the freedom to withdraw from the Inquiry if they disagreed with the KLMC's actions. The issue has always been the impacts and end uses of uranium mining, and the mandate the Inquiry was so "committed to discharge" was so questionable from the start that the decision to continue work at the mine site left no doubt about the Inquiry's outcome.

The final report stated, "We are satisfied on the information that we heard that these errors in administration were bona fide and that no improper purpose or motive was involved." But actual consequences, environmental harm in this case — not political and/bureaucratic motives, which are impossible to prove or disprove — should have been the Inquiry's concern. The overriding fact remains that the government-owned company illegally proceeded with Key Lake drainage and did so prior to the Inquiry being appointed. This reveals how uncommitted the government was to its own environmental assessment regulations. The Key Lake report was also in error in saying that the lawsuit brought against the mining company by the RGNNS regarding the illegal lake drainage was "subsequently dropped." By the time the case was heard in court, the judge ruled that the request for an injunction was unrealistic, in that the lake drainage was already 90% completed.

The Key Lake report did in fact recommend that the government not make future "mistakes in the administration of the law" that would jeopardize the credibility of "future inquiries." Such language calls into question the equality of the law and the distinction between "mistakes in law" and "breaking the law." When, for example, Métis and First Nations northerners make mistakes in law they are often jailed, even when their mistakes are victimless. But when the government breaks the law — with an assault on the environment, potentially with widespread victims — their actions are retroactively legalized. One way to avoid such political and legal risks would be to dispense with inquiries altogether. Indeed, soon after the KLBI report, the government decided not to have an independent inquiry for the new Gulf Minerals uranium mine at Collins Bay, near its existing Rabbit Lake mine.

The Key Lake final report tried to justify its pro-nuclear terms of reference by stating:

> It seemed reasonable to us that we were not asked to review again the larger question of nuclear development considering that the Cluff Lake Board had reported on that same question only 18 months before our appointment.

Two pages later, the report recommends that "future inquiries" deal with "issues or subjects which have not been adequately addressed or answered in previous Inquiries...." This raises a serious contradiction. How could the Key Lake inquiry fulfill that part of its mandate calling for a "review of all available information on the probable environmental, health, social and economic effects... of the mine and mill at Key Lake" without objectively examining the scientific, market and other information which had yet to come to light at the time of the Cluff Lake Board of Inquiry? Furthermore, a number of knowledgeable analysts have noted that assumptions of the CLBI, particularly that nuclear energy is needed, safe and economic, were never "adequately addressed."

In an attempt to strengthen the credibility of future inquiries, the KLBI report also recommended that in the future board members "should be selected with some particular expertise or perspective in mind." Such a recommendation was far too general and worked against the public good, for the Cluff Lake and Key Lake boards had been "selected with particular expertise or perspective in mind." Both were headed by a lawyer who brought a legalistic rather than a comparative, scientific perspective to bear. Rather than emphasizing careful deliberations about needs, causes and effects, the proceedings were dominated by rules, precedents and authority. While the technical expertise (e.g., nuclear chemistry, geology, etc.) was of a specialized nature, the overall ecological, energy policy and international perspectives were entirely absent.

Uranium Mining Chronology Tells It All

The historical chronology of uranium mining in Saskatchewan reveals how public inquiries have served the interests of Crown and private uranium corporations. The Saskatchewan Mining and Development Corporation (SMDC), the government's instrument in uranium exploration and mining, was created in 1974. The order-in-council giving it the power to own up to 50% of uranium joint ventures was approved in 1975. The EIS for the French uranium company Amok's Cluff Lake mine was initially approved by the Department of the Environment in 1976. All this occurred before there was any consideration of having an inquiry.

The CLBI was the outcome of an unexpected conflict over uranium mining at the 1976 NDP convention. At that time a party committee recommended a moratorium on uranium mining to allow a full public inquiry into the subject. Government officials and ministers, who had already committed their bureaucracies to uranium mining, particularly Jack Messer, Minister of Mineral Resources, fought successfully to have the moratorium defeated and a short inquiry was accepted as a "compromise." Although there was an attempt to present the CLBI as a fundamental examination of uranium mining, there was never any doubt that it would support industry-government initiatives already underway. Uranium markets were already committed, and Messer even publicly stated that the inquiry wouldn't alter this fact.

The Key Lake report simply perpetuated these shortcomings. There are fundamental gaps and errors in the report, which was based mostly on submissions by the chief proponent for expanded uranium mining, the KLMC. The report's assumptions about the economics of uranium mining, the social impact on the north, the health of workers, mining technology and the protection of the environment, especially uranium tailings storage, were never fundamentally questioned.

This raised serious doubt about the integrity of the environmental assessment procedure. Each inquiry appeared more and more political and less and less environmental. Judge Bayda, chair of the CLBI, went on to be become Chief Justice of the Saskatchewan Court of Appeal in 1981. Few were surprised to hear that in the same year Bob Mitchell, the Key Lake Board chair, announced his candidacy for the NDP in a Saskatoon constituency, and there was widespread speculation that, if elected to the seat, he would automatically go into Roy Romanow's cabinet. Later he became one of Romanow's most powerful ministers, acting as Minister of Justice for much of a term

Looking back, it seems inevitable that questions would be raised about whether there were ulterior motives behind the environmental assessment procedures used to expand uranium mining in Saskatchewan and whether the inquiries provided a credible foundation for the continuation of uranium mining and the expansion of nuclear power.

Impact of the Anti-Nuclear Boycott

The lack of credibility of the Key Lake Board of Inquiry, due to its terms of reference and the work already undertaken at the mine site, led to a full boycott by environmental and non-nuclear groups. Many of these groups had differed over participation in the earlier Cluff Lake and Warman uranium inquiries. The extent of this boycott, with *not one of these groups participating*, indicated the depth of the resistance to the inquiry process being used to legitimize the expansion of uranium mining.

Though the boycotting groups knew that their non-participation would be deliberately misrepresented as a sign of disinterest in the Key Lake mine, they also felt, considering that the mine was already under development, that their participation would allow the Inquiry to use them as non-nuclear "straw men" to justify the Inquiry's predetermined decision. It is true, as some argued, that participation could serve to reveal much of the corporate and government wheeling-and-dealing that had been going on behind the scenes, as the participation of Masie Shiell showed. With her series of articles in the *Catholic Weekly*, "Prairie Messenger," from December 1980 to March 1981, Masie became the elder conscience of the inquiry process and a thorn in the side of both government and industry. However, the participation of all or most of the non-nuclear groups would have helped restore the deteriorating credibility of the Inquiry. The public would have gotten a dual message from the non-nuclear groups, and their own credibility might have been jeopardized. The non-nuclear groups also realized that a broad-based alliance, even stronger than the one that opposed and stopped the uranium refinery at Warman, would have to be developed to successfully stop uranium mining.

At the same time as the KLBI was opening its formal hearings in La Ronge in June 1980, members of several Indigenous and non-Indigenous groups were marching through the town and the adjoining reserve to demonstrate their opposition to the Key Lake mine. After that there were two non-nuclear "Gatherings for Survival" at Pinehouse, on the road to Key Lake. These events included information workshops on such topics as the implications of uranium mining for Aboriginal rights.

The boycott therefore had the desired effect of shifting scarce voluntary resources from time-consuming ineffective participation in the inquiries to more effective grassroots organizing. This had not been without its challenges, for example, in developing a workable alliance between Indigenous and non-Indigenous people opposed to the mining.

The boycott also had a direct impact on the Inquiry itself. As the final report said: "The Board felt handicapped by their [non-nuclear groups] absence and, as a consequence, we found ourselves having to do a great deal more work." Because of the boycott, the bias in the Board's terms of reference, deliberations and final report were more evident than they would have been otherwise. The bias stood out clearly, free of the sorts of meandering obfuscations that had plagued the much larger Cluff Lake report, which had tried to give the impression that it was considering "all sides," when in fact it was constructing arguments to confirm its predetermined decision to expand uranium mining in the north. The much shorter Key Lake report was more easily assessed; it allowed more people to see the manner in which the government's inquiry process had been used to legitimize the ramming through of uranium mining.

Saskatchewan NDP Overturns Pro-Uranium Policy

But there were more problems in store for the nuclear proponents. The Blakeney NDP government had been re-elected in 1978 without having to justify its joint ventures in the Cluff Lake and Key Lake uranium mines. The 1982 provincial election, however, was a different matter. Non-nuclear and environmental groups were more vocal about their opposition to uranium mining; independent labour and Aboriginal candidates ran in some seats; and I ran as an "independent green" candidate in Regina Victoria, primarily to raise the issue of uranium mining and nuclear energy.

The opposition Tories cleverly exploited the growing dissent. Premier in waiting, Grant Devine, was regularly seen at the back of large public meetings where uranium mining was being criticized. Election pamphlets issued by the Progressive Conservative Party left the impression the party was critical of uranium mining. One stated: "The NDP government has invested $500 million in uranium. There is no return on that investment yet and the price of uranium has fallen by more than one-half. Investment in uranium does not help people today and it does not secure our future for tomorrow."

The NDP's cavalier defence of the Crown corporations involved in resource development was partly responsible for its devastating defeat at the polls. Now with only eight NDP MLAs as the Official Opposition, the non-nuclear grassroots of the party began to gain momentum. After a year of province-wide constituency educational meetings, the Saskatchewan NDP reversed its long-standing policy supporting uranium mining. By an overwhelming vote of 390 to 134 at its November 27, 1983, annual convention in Saskatoon, the party endorsed a motion to stop all expansion of the industry and to phase out existing uranium mines in the north.

The voting process was notably fairer than in previous conventions, although the debate remained predictable. The anti-uranium caucus handily won a compromise motion to phase out uranium mining, but there were early indications that the NDP leadership was not going to be committed to this change. Since the beginning of mining at Uranium City, during the nuclear arms race of the 1950s, the NDP government and its predecessor CCF governments had consistently supported uranium mining. The reversal of such a long-standing policy reflected the steady growth of the anti-nuclear movement and the sensitivity of politicians to growing anti-nuclear public opinion in the province. Though the successful compromise motion calling for a phase-out rather than a shutdown of existing mines had been the weakest of three anti-uranium motions submitted to the convention, at a previous NDP convention (when the NDP was still in government) the motion had received no more than 30% support.

Saskatchewan's new Tory premier and a spokesperson for the Crown's uranium corporation, the SMDC, both stated that this reversal of policy by the party that had masterminded the expansion of uranium mining in the 1970s, might lead potential investors to postpone major investment decisions until they saw how the electorate would vote in the provincial election expected in two to three years.

The anti-nuclear caucus in the NDP garnered an impressive list of supporters for the motion. Most Saskatchewan-based MPs and several MLAs supported the anti-uranium motion. (This included anti-nuke MLA Peter Prebble and Lorne Calvert, then environment critic, now pro-nuclear premier of Saskatchewan.) The motion was brought to the floor by a past senior cabinet minister, Wes Robbins, who argued that uranium should be kept in the ground until safe disposal methods were found and proven. It was seconded by party treasurer Al Hewitt, who, making the link between nuclear energy and armaments, said that he didn't want to have anything to do with uranium mining, primarily because of the cruise missile testing in northern Saskatchewan.

The procedures for the debate on the controversial motion were much improved compared to past NDP conventions. In her report to the 900-person convention, Betty Wiebe, who was one of the three anti-nuclear members of the six-person committee that arranged the pre-convention educationals, recommended that those for and against the motion alternate as speakers from the floor and be limited to three minutes. Wiebe also recommended that the question on the motion not be called until after forty-five minutes of floor debate and that the vote be a standing vote. These procedures were adopted, perhaps heading off the frequently used tactic of calling the question after several authority figures, sometimes including the party leader, had passionately spoken in defence of a status quo policy.

In contrast to the process, the content of the debate was little changed from past conventions. Pro-nuclear speakers claimed that the motion would undermine lifesaving nuclear medicine, that nuclear energy was less hazardous than other energy sources, that northerners and developing countries would be the victims of a shutdown and that it was not the technology itself but "political decisions" regarding its use that created the risks. Anti-nuclear speakers countered that the world's medical needs could be supplied by one vein of uranium; that you couldn't compare risks to individuals with the risks for huge populations from such threats as a nuclear meltdown; that the north had never been given real developmental alternatives or choices; and that environmentally appropriate technologies were required to create conditions for viable world peace as well as for sustainability.

Neither the pro- nor anti-nuclear speakers mentioned the role of the uranium policy in the NDP's over-all fiscal policy, particularly the emphasis on joint-ownership with several energy multinationals. Nor was there any debate on the implications for social or health policy of the previous $600 million of public moneys committed to uranium mining. It could be concluded from this that uranium mining was still seen as a single issue by large numbers within the NDP. For many, it was merely a mistaken policy, not the outcome of a mistaken approach to policy.

NDP ex-cabinet ministers who had defended uranium mining vehemently while in government were noticeably absent at the convention microphones. The responsibility to defend past policies and practices was left to Terry Stevens, a representative of the United Steelworkers of America; Bob Mitchell, a past inquiry chair; Angus Groome, a past commissioner from an NDP government-appointed uranium inquiry; and Dale Schmeicel, a past public relations officer with the NDP-established SMDC, who was to become editor of the NDP paper, *The Commonwealth*. Party leader Allan Blakeney appeared at the pro-nuclear microphone just as the time allotted for debate ended. He did not support the motion to extend the debate, which was defeated; thus the party establishment made only a symbolic gesture in defence of a policy that by then was clearly not going to be upheld by the grassroots of the party.

Allan Blakeney opposed the anti-uranium motion during the standing vote. The bulk of the opposition to the phase-out of uranium mining came from past government technocrats, several union delegates (who some have argued did not fairly represent the sentiments of the union membership in the Saskatchewan Federation of Labour) and a few northern Indigenous NDP politicians. Support for the motion came from peace and environmental activists and the vast majority of women and youth delegates to the convention.

Symbolic or Real Paradigm Shift?

These differing pro- and anti-nuclear alliances reflected broader political alignments and re-alignments. The growing ecology, feminist and disarmament movements throughout much of the world were at the time beginning to make vital analytical connections. They were evolving a new vision of civil society and social development that was very different from the economic-growth, trickle-down ideology advanced by patriarchal elites in the corporations, the dominant political parties and the labour and social democratic establishment. While this underlying realignment clearly influenced a change in NDP uranium policy, it was doubtful that the alternative development path would find any fundamental support within the NDP without substantial reorganization of the party, its structure and leadership. Even though they received only 25% of the vote, party members associated with the pro-nuclear position maintained major control of the party's leadership (including electing several vice-presidents) and of party organization.

At his post-convention press conference Blakeney carefully managed the call for "renewal," which is typical of the NDP when they are out of power, stressing the dependence of the party's grassroots on the technocrats to regain power and manage the affairs of government. Blakeney was quick to emphasize that, because the motion called for a responsible phase-out of uranium mining, nothing immediate would happen if the NDP were re-elected, suggesting that the victory of the grassroots was "symbolic" only.

But Blakeney did concede, for the first time, that few royalties come from uranium, meaning that the impact of a phase-out on the provincial treasury would be minimal. This amounted to an admission of defeat for a policy Blakeney had carefully stage-managed as part of a strategy to diversify Saskatchewan's economy and to hold power in the 1978 election. In 1977 the NDP government had predicted that, by 1983, uranium revenues would be between $112 and $224 million, but the actual figure was only $29 million.[2] Blakeney failed, however, to mention the drain on the province from the massive allocation of resources into uranium mining in the false hope of windfall profits and revenues for his government. Nor did he mention the steady accumulation of uranium tailings resulting from his government's policies.

These developments in uranium mining policy brought the Saskatchewan NDP back into step with the federal NDP on this issue. With the shift in policy some previous party members who had been disillusioned by the NDP's "business as usual" approach to uranium mining rejoined the party. Their hope that this was a turning point for the Saskatchewan NDP was soon to be quashed. The overturning of the pro-uranium policy also challenged those in the province committed to alternative green politics to be more thoroughgoing in their analysis of the limitations and pitfalls of social democracy.

Spilling the Beans on the Key Lake Spill

As if to add insult to injury, the image of a responsible regulated expansion of uranium mining, which the Saskatchewan NDP government wanted to project in the late 1970s, was soon to be fundamentally tainted. A massive radioactive spill at the Key Lake mine occurred only a few months after this "state of the art" mine opened. Realizing the serious implications for the credibility of the uranium industry, the Saskatchewan Mining Association (SMA) quickly went into damage control mode. The energy alternatives proposed by non-nuclear groups were soon to be taken more seriously as a broader spectrum of the public began considering the long-term environmental and health risks at the front-end of the nuclear system.

A critical look at the chronology of the Key Lake mine and how the threat of spills was handled in the KLBI reveals once again that inquiries were being used primarily as a tool for managing political dissent rather than a process to examine the fundamental issues. Most disconcerting is the evidence that the KLBI simply paraphrased the Key Lake Mining Corporation submissions in significant areas of its own final report. This not only sheds light on the dangers of collusion between corporate and government bodies but is testimony to the quality of governance in pro-nuclear administrations.

The Key Lake uranium mine in northern Saskatchewan began operating in October 1983. Within three months there had already been eight spills totalling over 1.5 million litres of radioactive liquid waste. The largest spill, of 100 million litres on January 1, 1984, brought the mine to the attention of the national media. Coverage focused narrowly on whether the spills constituted an *immediate* hazard. The underlying issues over standards, methods, evidence and time-span were not explored. Nor was there an investigation of the more fundamental problem of the total accumulation of wastes (liquid, solid, liquid tailings and gas evaporation). Using the company's own 1979 figures, there would be nearly 60 *billion* pounds of total waste if this mine were allowed to operate for fifteen years.

Industry-related experts predictably downplayed the spill. Federal and provincial regulatory bodies repeatedly contradicted each other. Environmental and non-nuclear groups, who had observed the inquiry process and were skeptical beforehand, responded with "we told you so!" Peace, labour and northern groups became more critical of the uranium industry as a result of the spill. And the SMA began an expensive public relations campaign to try to cut their losses in credibility for the uranium industry.

A Spill That "Could Never Happen"

A brief history of the fiascos associated with the Key Lake mine helps put the spill in context. When submitting the company's EIS in October 1979, KLMC president Peter Clarke assured the public that "the project is designed to be built and operated with the least possible disruption to the environment." This view was also expressed in the company's 1979 promotional document, *The Key Lake Mine Project*:

> The tailings left after the uranium is extracted will be piped to the tailings management area. Water used in the mill process will be either recycled for further use in the mill or treated and tested to ensure that it meets applicable water quality standards before it is released to the environment.... This design provides, therefore, for the total containment of solid wastes, and control and treatment of all liquid wastes from the mining and milling operations throughout the operating life of the project and after decommissioning.

A nearly identical statement is found in the Board of Inquiry's summary report, issued two years later, in 1981. Of special interest, in view of the 1984 revelations about a series of liquid spills, is the removal of the phrase: "and control and treatment of all liquid wastes from the mining and milling operations." Why was this reference to the liquid waste system deleted? Did the Inquiry have pre-cognition? Did it willingly suspend critical judgement or simply overlook the critical issue of liquid waste? In his public statement, which I heard on the local CBC radio, about the 100 million litre Key Lake spill, Inquiry chair Bob Mitchell said, incredulously: "This just could not have happened based on the presentation that the KLMC made in 1980. They proved to our complete satisfaction that what happened could not have happened."

Even his public display of surprise was questionable. Volume 25 of the KLBI transcripts revealed that an Alberta hydrologist, brought in by the Northern Municipal Council (NMC), had questioned the company's assumptions about the amount of water that would have to be contained (the ore is under water) and the ability of the design to handle this and other liquid wastes. This information, especially in the context of the Inquiry directly plagiarizing the company's report, raises the question of the Board of Inquiry's complicity in the spills.

Pages 3 to 7 of the KLBI summary report are lifted nearly verbatim from pages 7 to 19 of the earlier KLMC promotional report. The position taken by the KLBI on the surrounding geography, yellowcake, the ore bodies, the mine design, the lakes, the over-burden, groundwater, mining and milling and waste management is nearly identical to that of the uranium corporation. One of the more blatant examples of this peculiar coincidence concerns the draining of the lakes over the ore bodies. The company's promotional report, published in 1979 before the public controversy over the lake drainage, and the Inquiry report, published in 1981 after the controversy, contain the exact same statement: "In order to permit an assessment of the project's feasibility, it was necessary to lower the level of Key Lake and seven smaller lakes. Canals were dug in order to divert the flow of surface water from the area of future mine production. This work has been carried out under the authority of government approvals relating to the investigational stage of the project." This is a falsified depiction of the circumstances surrounding the drainage.

At the time of the major spill, media attention focused on Peter Clarke, KLMC president. Clarke played the front man, taking the flak for the industry. The media did not explore the multinational or federal and provincial bodies associated with the mine. Nor did they focus on the dual role of the Saskatchewan government, which was part owner and which had to appear to regulate the industry. The provincial government's record had thus far been atrocious. Under the conditions of the province's Surface Lease Agreement, the Northern Monitoring Committee was to review the inspection reports of the Department of Environment as well as the company's hiring policies and the mine's socio-economic impact. At the time this Committee had not met for over a year. Under the lease agreement the Department of Labour was to have trained on-site workers in mine monitoring. This had not been done. When the NDP government issued this lease, it said it would be tough in its enforcement. But even while out of office, when the Key Lake spills occurred, the NDP did not call for a mine shutdown even though the conditions of the original lease had not been met.

Another owner of the Key Lake mine was Eldor, a subsidiary of Eldorado Nuclear, a federal Crown corporation. The federal regulatory body, the Atomic Energy Control Board (AECB), dominated by nuclear industry insiders, was under scrutiny because a board member was also president of a company with large sales to Ontario Hydro, the largest nuclear power corporation in Canada. Not surprisingly, the AECB, contradicting the provincial Department of Environment, had recently declared that the Key Lake company did not have to clean up the January spill. As the public controversy over the spill intensified, some public advocates, including the Inter-Church Uranium Committee (ICUC), called for charges to be laid against the company. There were also calls for a public inquiry from the ICUC, the Northern Rights Coalition and even the provincial NDP caucus. Others, having little faith in past government inquiries, called for an immediate shutdown of the mine. Advocates of this included the Regina Coalition for Peace and Disarmament and the Regina Group for a Non-Nuclear Society.

Key Lake Go-ahead before Cluff Lake Inquiry

As outlined earlier in this chapter, the KLBI was steeped in controversy from its beginning. Because the company had already drained several lakes near the mine site well before the KLBI was appointed, its credibility was in question. Furthermore, it "was specifically not requested to review the broader implications of uranium mining and milling and the nuclear industry" because the CLBI had supposedly addressed and resolved these broader issues. But even leaving aside the evident biases of the Cluff Lake Board of Inquiry, this justification collapses — since the Key Lake mine was underway before the CLBI report was even completed.

The chronology of events is most revealing. On April 28, 1977, the German firm Uranerz, one of the original Key Lake companies, applied to the Department of Mineral Resources to de-water lakes near the Gaertner and Deilmann (Key Lake) ore bodies. *The CLBI had only been appointed two months before this.* This request was approved by Minister Jack Messer on June 2, 1977. On September 27, 1977, Uranerz applied to construct the de-watering canals. This was also approved by Messer, on October 27, 1977.

On February 23, 1978, Uranerz applied to discharge "industrial waste" coming from the de-watering. Messer approved this on May 4, 1978, *a month before the CLBI had even finished its final report.* By July the surface drainage was underway. The public did not become aware of the lake drainage until government memos were leaked in June 1979.[3]

In August 1979, the RGNNS tried to stop the lake drainage through the courts, arguing that the mineral resource permits were not legal since the actions should have been approved under the *Water Resources Act*, by the Department of Environment. Though the court action failed, primarily because the drainage was already 90% completed, these revelations on the Key Lake mine were a turning point in raising the public profile of the uranium mine controversy in Saskatchewan.

That these events occurred five months before the KLBI was even appointed on December 11, 1979, clearly demonstrates that Saskatchewan's uranium inquiries were effectively nothing but afterthoughts — a means to manage public dissent over uranium mining rather than a serious exercise in public policy-making and participation.

The story continues. On February 2, 1980, in an attempt to restore some of its credibility, the Inquiry put a freeze on on-site work at Key Lake. This "show of strength," however, was short-lived. On July 16, during its formal hearings, the Inquiry lifted the freeze due to what it called "dire circumstances" facing the company. These turned out to be worries that the company would not be able to meet its 1983 start-up date.

The Inquiry fulfilled this ritualistic exercise and submitted its final report January 12, 1981. In February the NDP government gave its formal go-ahead to the project, which had been going ahead the entire time. This political double-think reached even greater heights: In June 1981 the company's de-watering lease expired but it continued draining the lake. This time the company was caught — but the response of Roy Romano, Attorney General, was not to lay charges but to make a new lease granted in September retroactive to June.

On October 1981, under growing public pressure, Ted Bowerman, the Environment Minister at the time, talked tough in Saskatchewan's big-city daily papers about establishing an orderly and comprehensive process that must and will be followed. If this is not possible, he added, then there will not be, cannot be, a uranium mining industry in Saskatchewan.[4] When the Ministry finally laid charges, the company was fined the grand sum of $500 — for not abiding by the *Water Resources Act*. Feigning that this penalty would somehow end the trail of fiascos and appease the critics, Bowerman said, apparently with a straight face: "I suspect that it will be noticed by the industry."

The massive spills at Key Lake in 1984 proved the NDP minister dead wrong. They showed that both the uranium industry and the government's uranium inquiries lacked fundamental credibility. And because of these spills there was deepening support in Saskatchewan to stop the nuclear industry "at the source."

Notes

1. Key Lake Board of Inquiry, *Final Report*, Jan. 12, 1981, cover letter. All quotes from the KLBI are from this report.
2. See Jim Harding (ed.), *Social Policy and Social Justice*, Wilfrid Laurier University Press, 1995, p. 347.
3. See *Briarpatch*, Oct. 1979.
4. This and other events having to do with the Key Lake mine and spill are chronicled in "The Gulliver Key Lake Mining Corporation Dossier," available at <www.sea-us.org.au/gulliver/keylake.html> accessed July 2007. Also see Roger Moody, *The Gulliver Files: Mines, People and Land — A Global Battleground*, London, Minewatch, 1992.

chapter 5

CORPORATE AGENDA

From the late 1970s to the early 1980s those of us active in Saskatchewan's fledgling non-nuclear movement were often left reacting to the government's agenda and time-line. There were three inquiries in as many years, for two massive new mines and a proposed uranium refinery, all of which took their toll. After some of the pressure subsided, non-nuclear activists turned to researching the structure of the uranium industry. The discovery of the Cigar Lake high-grade ore body in 1983 had brought even more multinationals, from more nuclear countries, into the province. Japan, as well as France and the United States, now looked to northern Saskatchewan for cheap uranium fuel, and the province was now unquestionably the front-end supplier of the global nuclear system.

Cigar Lake Discovery: In the Eye of the Nuclear Storm

When the Key Lake uranium deposit was discovered in 1975 it was described as a "monster" find, and the mine was to become the largest in the world. But after 1983, attention shifted to Cigar Lake, about 115 kms to the north-east. According to the SMDC, this newly found deposit was even larger than at Key Lake, and to this day it remains the second largest high-grade uranium deposit in the world.[1] The largest high-grade deposit is also in northern Saskatchewan, at McArthur River.

The only other country known to have such large uranium deposits at the time was Australia, and although its deposits are estimated at more than twice the size of Saskatchewan's, the grade of Saskatchewan uranium is much higher. For example, though the 2% grade uranium at Key Lake is only one-quarter that at Cigar Lake, it is ten times the grade of Australia's Ranger mine and six times the grade at Australia's Jabiluka mine. More recently, large deposits of low-grade ore have also been located in Kazakhstan.

Economic considerations are uppermost in decisions about proceeding with such mines. When the projected demand for, and price of, uranium fell dramatically in 1977, Saskatchewan's first uranium mines near Uranium City closed and several companies announced they were abandoning exploration activities. These developments, occurring before the Key Lake mine began production, prompted a shift in investments to the higher-grade ore sites in the hope of lowering costs, capturing new markets and maintaining desired profit rates.

Such considerations about market and profitability placed Saskatchewan at the centre of the global uranium mining controversy. Due to the lower grades of ore and a national Labour Party ban on new uranium mines in Australia (with the exception of Olympic Dam), Saskatchewan deposits became much more attractive to the uranium multinationals and user countries. There was the added advantage that Saskatchewan uranium could be transported more easily to the big four users of nuclear-generated electricity: France, Japan, the United States and the now defunct Soviet Union.

The Cigar Lake project was expected to commence production by 1993; however, it wasn't even licensed until 2004. Unlike the Key Lake operation, which is an open pit mine, the Cigar Lake project is an underground mine, with the uranium ore over 400 metres below ground. The combination of exceptionally high-grade ore and underground mining presents serious engineering, production cost and radiation protection challenges. By keeping the corporation on guard, the Greenpeace-supported campaign against Cigar Lake may also have helped slow down the production schedule, providing valuable time for renewable energy alternatives to gain ground, which can be considered a small victory for the non-nuclear movement.

The companies owning the Cigar Lake site have, however, not been deterred. An unnamed president of a uranium mining company operating in Saskatchewan is reported to have said: "I'd rather face the technical problems of mining Cigar Lake than the political hassle of developing a uranium mine in Australia." He was probably referring to the fact that until recently in Australia, unlike Saskatchewan, Aboriginal rights have played a role in whether uranium mines proceed. Aboriginal rights were excluded from Saskatchewan's CLBI terms of reference, whereas the Aboriginal Land Councils in northern Australia had short-term veto power over uranium exploration at the time, and operating mines have revenue-sharing agreements with Aboriginal communities.[2] This shows the importance of effective non-nuclear alliances over Aboriginal rights and environmental protection in shaping investment decisions.

Cameco Dominant in Waterbury Lake Joint Venture

The firms involved in the Cigar Lake project were first called the Waterbury Lake Joint Venture (WLJV) — now called the Cigar Lake Joint Venture (CLJV). The firms initially included Cogema Canada of Montreal, a subsidiary of Compagnie Generale des Matieres Nucleares of France, which was controlled by the Commissariat de l'Energie Atomique (CEA) and which owned 37% of the joint venture. In 1983 Cogema became co-shareholder with the French corporation Amok, based in Saskatoon, which owned 80% of Cluff mining and, like Cogema, was tied into the CEA. In 2001 the French government created a single holding company for all its uranium and nuclear Crown corporations, named "Areva" after a Spanish Cistercian abbey, which ironically means "symmetry and dignity." It combined the CEA, the uranium company Cogema and the nuclear power company Framatome. This integrated French uranium/nuclear corporation now has bilateral agreements with the U.S., which is considering expanding nuclear power. Areva (Cogema) maintains 37% control of the Cigar Lake venture.

The CLJV also includes Idemitsu Uranium, based in Calgary, which owns the smallest share of the deposit (initially 12% and now 8%) and is a subsidiary of Idemitsu Kosan, Japan's largest petroleum company — an indication of the growing link between the petroleum and nuclear industries. This brought all of the world's largest producers of nuclear power directly into Saskatchewan's uranium economy and provided more impetus for local non-nuclear groups to network with Japanese groups.

The smallest owner is now Tepco (Tokyo Electrical Power Company) of Japan, which owns 5% of the CLJV, which it bought from Idemitsu. The largest original owner of the Cigar Lake venture was the SMDC, privatized in 1988 as Cameco, which maintains the controlling interest, 50% of the joint venture.

The French Connection: Uranium Still Going for Weapons?

As our research on the corporate connections deepened, it became clear that Saskatchewan was not only becoming the "Saudi Arabia of uranium" but was directly integrated into the nuclear weapons system. This, even more than earlier environmental health concerns, brought new people into the "community of action." With its seemingly righteous CCF-NDP heritage, Saskatchewan was now directly confronted by the moral challenges of its link with nuclear weapons.

The French connection was structurally and politically the most significant. At the time of the Cigar Lake discovery, France was ruled by a "socialist" government, not unlike the former "social democratic" government that expanded the uranium industry in Saskatchewan. Both governments, like the Canadian government itself, were deeply involved in ownership as well as regulation of the uranium and/or nuclear industries.

Initially the CLJV involved two state-owned corporations: the SMDC in Saskatchewan and the French uranium company Cogema. Cogema, the only non-communist Crown corporation connected to the entire nuclear fuel system, was linked, via the French CEA, to nuclear weapons' production and testing. If anything, the new French super-company Areva represents an even more integrated military-industrial nuclear corporation. As the largest nuclear power company in the world, it is facing the problem of accumulating nuclear wastes. Its mining arm accounts for 25% of the world's uranium production, making it and Saskatchewan-based Cameco, until recently, the two largest uranium companies and uranium tailing polluters on the planet.

France is second only to the U.S. in its nuclear-electrical capacity and is one of the major nuclear weapons powers, so the French connection clearly ties Saskatchewan directly into the global nuclear system. After discovering this, Saskatchewan non-nuclear groups began to make links with the French anti-nuclear and environmental movements, including Les Vertes and Parti Ecologiste.

With new high-grade ore deposit discoveries and mounting multinational corporate involvement in northern Saskatchewan, critical research and direct action about uranium mining in Saskatchewan achieved global significance. At the same time, our sister province, Manitoba, was being considered as a nuclear spent fuel storage site.[3] Cruise missile testing was being practised in our other adjoining province, Alberta, and the U.S. had just made the decision to locate armed cruise missiles in North Dakota, just south of Saskatchewan. The Canadian Prairie was becoming industrialized and militarized with nuclear technology, and Saskatchewan was quickly moving to the centre of the global nuclear controversy.

A 2007 merger of Canadian-based Uranium One with South African-based URAsia Energy made this company the world's second largest uranium conglomerate after Cameco. It controls large uranium reserves in both Kazakhstan and South Africa, and is behind Australia's Honeymoon uranium mine. Uranium One hopes to exploit the shortfall of world uranium production expected from a further postponement of the Cigar Lake start-up to 2008, after a major flood of the underground tunnels in 2006.

Comeback Strategy of the Canadian Nuclear Industry

While uranium multinationals were swarming into northern Saskatchewan, the Canadian nuclear industry was laying plans to piggyback the "uranium boom" with its own expansion strategy. In September 1987, the Canadian Nuclear Association (CNA) launched a multi-million-dollar, cross-Canada, promotions campaign. Despite the onslaught of TV, radio and print ads, I did not know the underlying agenda until 1990, when I got a leaked copy of the "CNA Public Information Program: 1987–1988 Business Plan," circulated by the AECL.

The AECL — the government's Crown corporation in the business of selling the Candu and other nuclear technology — was facing desperate times and needed a front group to help promote it. The AECL's September 22, 1987, corporate office cover letter to the CNA document revealed that the "AECL has agreed to pay $2.5M towards improving public acceptance of nuclear power, this to be done under the auspices of the CNA." The Business Plan shows that the AECL and CNA objectives were two-fold: to do a mass media campaign, while at the same time infiltrating mainstream institutions.

In the 1970s the nuclear proponents, with their futuristic scenario of an Electric Society, predicted a massive increase in demand for nuclear reactors as a result of the apparent insecurity of fossil fuel supply and the alleged impracticality of alternative fuels and energy sources. However the major demand for uranium fuel, especially during the Reagan presidency, was not for reactors but from the military. Throughout the renewed nuclear arms race of the 1980s the commercial uranium and nuclear industries continued to claim that they were not connected to the military. However, the AECL had not sold a single Candu reactor in over a decade, until its 1990 sale to South Korea, a decade when, according to their experts, the shift to nuclear energy would occur. The so-called "peaceful" nuclear industry had become increasingly dependent upon the military. The federal Energy, Mines and Resources Standing Committee starkly admitted on page 145 of its August 1988 report, *Nuclear Energy: Unmasking the Mystery*, that without government subsides (even if reactor sales were made), the survival of the nuclear industry would depend upon contracts with the nuclear submarine program proposed by the Mulroney federal government.[4]

The AECL knew it couldn't survive in the market if it was fully privatized, so the CNA went on the offensive to animate business and public support for a few reactor sales, making the AECL seem like a viable industry to its main backer, the federal government. In its 1987–1988 promotional campaign, the CNA relied on no-holds-barred tactics, using millions of dollars from the public purse, with Saskatchewan's uranium mining industry and pro-nuclear elites as its best allies.

In order to save the AECL's neck, the CNA engaged in overt and covert manipulation. In the name of providing "information" to a skeptical public and to refurbish its image as an open and accessible industry, the CNA planned one of the most cynical disinformation campaigns in Canadian history. As the leaked Business Plan showed, the nuclear industry broke new ground in the perpetuation of deceit.

The overall goal of the CNA campaign was to make Canadians "more aware of the benefits to be derived from peaceful applications of nuclear technology." This ruled out any pretence at balanced public education, as that would mean also paying attention to the burdens of nuclear energy[5] and exploring the question of whether the so-called "peaceful atom" was truly separate from the "military atom." Furthermore, the nuclear industry would have to be assessed in relation to alternative scenarios (energy futures), which assumed an equitable allocation of subsidies, as well as an equal capacity to inform the public of such options. But there was no more of a "level playing field" then than there is now. While the federal Conservative government was phasing out subsidies for alternative energy technologies in the 1980s, it continued to provide $140 to $200 million yearly to the AECL.

Through careful semantics the industry presented itself as objective and open, while its ends were entirely self-serving. The CNA's "key role" remained enhancing the community's "acceptance of nuclear technology," and the "understanding" it desired was always directed towards this end. Allowing the public a truly comprehensive understanding of the burdens of nuclear power and its alternatives would inevitably reduce support for the industry and was therefore not an acceptable strategy.

It is warranted to look at the industry's information campaign in terms of the tenets of propaganda. To persuade Canadians to accept the nuclear industry the CNA positioned itself between its funders and a number of "crucial publics." The leaked CNA Business Plan states:

> The number of publics whose views the CNA must consider and with whom the CNA must communicate will increase as people coalesce around matters of common interest on a more local basis. At the same time, interest groups are becoming more sophisticated. The result will be a more complex public affairs environment and the need for increased specialization in communications.
>
> The tendency for individuals or groups to focus on single issues will make it more difficult for the industry to address public concerns. Information gathering and integration will become more complex. Similarly, it will become more difficult to define public acceptability issues and respond to them collectively, when decisions are made....
>
> The CNA will need to continue to gather data about the external environment, to listen to public inputs, and to interpret this information on behalf of the industry. It will have to anticipate significant issues effectively, both in the short and long term, and take a strong leadership role in recommending action. The timeliness of the Information Program group's input to members will be an important element to industry performance.

This analysis, interpretation and strategizing "on behalf of the industry" was clearly an attempt to marshal corporate resources and political and economic forces in order to weaken opposition and to alter the negative public profile of the industry. The overall aim, in its own words, was "directed at developing a strategic framework for industry input, including an issue management program." This included "preparations for involvement in government decision-making with respect to waste management," an issue on which the industry knew it was particularly vulnerable. The bottom line was not public education and understanding but "expanding industry effectiveness and penetration." The strategy was to make corporate communications appear purely informational, thereby enhancing the industry's credibility — maneuvers reminiscent of non-democratic propaganda campaigns and a threat to the autonomy of public and democratic institutions.

The CNA's own program priorities and resources expose the educational pretence of their campaign. Of the twelve programs identified as part of the Business Plan, only one — waste management — directly involved a substantive issue rooted in the scientific and ecological controversy about nuclear technology. And this was selected as a program because it was seen as an issue that needed to be publicly managed in order for the overall program to succeed, as recurring polling had shown that the industry's credibility was particularly weak because of a lack of public trust that a solution to the nuclear waste problem existed. If concern for nuclear waste had been sincere, the CNA would also have targeted uranium mine tailings, one of the greatest waste problems of the overall nuclear industry.

Other CNA programs could be categorized as information, dissemination and fronting. Information included research, print material, videos, TV commercials and an advisory council. Dissemination included mass advertising and an information-phone line. Fronting included support groups working in the schools and media and "public participation" tactics. The allocation of resources within these areas is very revealing. Of the nearly $7 million budgeted to be spent in 1987 and 1988, according to the leaked Business Plan, over $4.5 million, or two-thirds, was spent on advertising.

Over one million dollars went into motivational and communications' research for the CNA information packages. The rest went into creating and mobilizing "natural allies" within the scientific, education and media organizations. The largest amount, $200,000, was allocated to create support groups, most of it likely going to Saskatchewan, soon to be targeted for a Candu reactor. The next largest amount, $91,000, went to penetrating Canadian schools, where the nuclear industry could disseminate its costly and slick print and video propaganda to teachers and students. These materials flooded Saskatchewan, from schools to malls. The smallest amount, $16,500, was used to create an image of public participation, though active and critical public participation wasn't encouraged. Almost three times this amount, $50,000, went to maintain the centralized and controlled telephone information line promoting nuclear power.

Those who created the CNA's Business Plan didn't think it would cost much (only $30,000) to get pro-nuclear scientists onside in an advisory capacity. The convergence of interests (including financial) among nuclear scientists, technicians and engineers is well known, and the use of public resources within their own institutions to aid "their" industry would add additional backing for the CNA campaign. In chapter 9, I discuss how some faculty members at the University of Regina supported the push to nuclear power, seemingly with the full support of senior administration.

Particularly offensive was the penetration of the public school system and the "free" press, which has major implications for the credibility of these mainstream institutions. The tactics used in this regard were an assault on the spirit of equity and fair play fundamental to any pluralistic society and are of such importance that I discuss them at length in chapter 15.

The cynical use of clandestine tactics by the nuclear industry is demonstrated by the fact that, while this CNA "public information" campaign was occurring, the AECL was creating a secret database on the personnel, funding and vulnerabilities of about a hundred anti-nuclear groups across Canada. This "Database on Anti-Nuclear Groups" was published for "Restricted Commercial" use by the AECL in January 12, 1988. It was done by the Ridley Research Group of Toronto and was leaked to several anti-nuclear activists across Canada including myself. As this nuclear offensive continued, it felt like the non-nuclear forces were on the receiving end of a counter-insurgency campaign to push back the gains made over the previous decade. Shutting down the nuclear industry was not going to get any easier.

Notes

1. The largest high-grade deposit, found at McArthur River, of which Cameco owns 56%, is also in northern Saskatchewan. See <http://www.cameco.com/operations/uranium/mcarthur_river/> accessed June 2007.
2. Jim Harding, *Aboriginal Rights and Government Wrongs: Uranium Mining and Neo-colonialism in Northern Saskatchewan*, In The Public Interest: Working Paper No. 1, Prairie Justice Research, University of Regina, 1992.
3. See Walter Robbins, *Getting The Shaft: The Radioactive Waste Controversy in Manitoba*, Queenston House, 1984.
4. When the Chrétien Liberals quashed the nuclear submarine deal after being elected in 1993, the CNA and its member organizations were left behind the eight ball.
5. See Jim Harding, "The Burdens and Benefits of Growth: Mineral Resource Revenues and Heritage Fund allocations under the Sask. NDP, 1971–82," in Jim Harding (ed.), *Social Policy and Social Justice*, Wilfrid Laurier University Press, 1995, pp 341–74.

chapter 6

FREE TRADE AND URANIUM

During the 1988 federal election, fought over the Mulroney government's U.S.-Canada Free Trade Agreement (FTA), there was general talk of the implications of the trade deal for Canadian energy resources, but there was no attention paid to uranium specifically. Meanwhile, the Saskatchewan uranium industry was hoping to benefit from the free trade deal and became one of its earliest proponents.

Saskatchewan's corporate media was also solidly onside. On November 18, 1988, the Regina *Leader Post* published a short version of my concerns about uranium and free trade, which was juxtaposed with a pro-FTA piece by SMDC head Roy Lloyd. The *Leader Post* gave the impression that they invited me to express my views, implying a balanced editorial approach, but the fact is that I was not asked. I pressured the paper to provide some balance to its blatantly pro-FTA coverage, and much of my argument was gutted from the final piece.

At the time, the U.S. was facing the high cost of developing its remaining low-grade uranium reserves as well as rising costs from mine decommissioning and uranium worker compensation claims. Saskatchewan's high-grade uranium was seen as a reliable, strategic supply, and free trade was therefore critical to the U.S. for its continued military-industrial supremacy.

This aspect of the FTA was ignored or minimized by Canadian groups opposing the deal. Although the Saskatchewan NDP occasionally opposed the FTA, especially regarding agriculture, it wasn't about to expose its "golden egg," uranium mining, to more public scrutiny. Prior to the International Uranium Congress (IUC), held in Saskatoon in June 1988, congress organizers attempted to ally with the high-profile group, Citizens Against Free Trade, headed by David Orchard, but uranium mining was not then on its radar screen. The important connection between the free trade debate and uranium mining was still to be made.

Impact on Saskatchewan

It should come as no surprise that Saskatchewan is directly tied into free trade since northern Saskatchewan is now the main area of uranium mining in Canada, and Canada is now the world's largest uranium producer. Given the power of the uranium industry and the SMDC's early enthusiastic support of the FTA, what is surprising is that there hasn't been more debate about the matter. This lack of debate may be intentional, as all governments, including several Saskatchewan NDP governments, support expanding uranium exports to the U.S.

Concerted lobbying over the years ensured that the FTA dealt with uranium mining. The role of one man, Donald McDonald, stands out. He was involved in the uranium cartel's price fixing in the 1970s and also headed the Royal Commission on Economic Union and Development Prospects, which recommended free trade with the U.S. After the Mulroney government released the FTA text, McDonald co-chaired the Alliance for Trade and Job Opportunities, a corporate lobbyist for free trade. This is the same Donald McDonald who, as a contender for the Liberal leadership after Trudeau resigned, suggested that Canada build a string of nuclear power plants along the 49th parallel to help increase exports of electricity to the U.S. His vision for Canada, like the FTA's, amounted to Canada being a supply station for the U.S.

Uranium exports were used as an important reason to support the FTA because, like the Auto Pact, they strengthen Canada's trade surplus with the U.S. From the 1950s to 1960s most of Canada's uranium has been exported to the U.S. for nuclear weapons and electrical production. Eighty percent continues to be exported in 2007, half of it to the U.S. Although the value of the sales is high, it is much lower than agriculture, forestry or oil and gas exports.

More important than the value of exports is the distribution of the wealth and the social and environmental costs of continued mining. According to the province's annual *Economic Reviews*, between 1975 and 1985, the value of Saskatchewan uranium sales was around $2 billion. Yet, in spite of the boom promised by past NDP governments, according to Saskatchewan's *Public Accounts*, during that period only $130 million came back to the province as revenues. This was far less than was spent by the province during the period to expand its uranium Crown corporation's ventures.

The uranium industry's argument that exports create jobs was the same one used by FTA proponents, but, as in all resource industries, capital and technology are replacing jobs in uranium mining. The industry produced only half the direct jobs that were promised at the Cluff Lake Board of Inquiry, and very few of these, about 15%, went to northern Native people.[1] Northern unemployment did not lessen as the uranium industry expanded, and the long-term environmental and social costs are being left for the people of the province and the north to bear. Concerned about its own bottom line, the uranium industry has none of these concerns as its priority.

The FTA Text on Uranium

FTA supporters claimed critics misrepresented the text, so it is helpful to look at the extensive *Explanatory Notes*, published by the federal government in 1988 on the "U.S.-Canada Free Trade Agreement." The text exposes the generalities about there being a "common interest in ensuring access to each other's markets and enhancing their mutual security of supply." The meaning of "common interest" has to be severely stretched if we are to believe that this is a win/win deal between equal trading partners. Within the legal text, the U.S. is clearly the importer and Canada the exporter. While Canada has "access to the U.S. market for energy goods," it is the U.S. that gets security of supply, i.e., access to Canadian resources.

The FTA further institutionalized this dependency relationship with the U.S., shown in the section on "Energy," which stated: "The U.S. has agreed to eliminate all U.S. restrictions on the enrichment of Canadian uranium and Canada will eliminate the requirement for uranium to be processed before it is exported to the U.S." The concrete implications of this "trade-off" for the two countries is shown in Annex 902.5, which states, under "import measures:" "The U.S. shall exempt Canada from any restriction on the enrichment of foreign uranium under section 161V of the U.S. Atomic Energy Act." Under "export measures," it states: "Canada shall exempt the U.S. from the Canadian Uranium Upgrading Policy announced by the Minister of State for Mines on October 18, 1985."

The above clauses indicate that, under the FTA, the U.S. could enrich "foreign" uranium from Canada for its own use, without restriction, if Canada no longer requires that its uranium be upgraded (i.e., refined in Ontario) before export. But this is slightly silly as well as deceptive. The U.S. wants and needs high-grade, low-priced Canadian uranium, which must be enriched for use in the light water reactor design in the U.S. or for U.S. weapons' production. Furthermore, they insist on enriching it themselves; in any case, Canada has no enrichment facilities, as enriched uranium is not required in the Candu reactor. So not only did the FTA give the U.S. security of Canadian uranium supply, it also gave them control over processing, refining and all enrichment.

Under the FTA, Canadian uranium became treated as having its "domestic origin" in the U.S. This isn't because the U.S. thinks highly of Canada as its major trading partner, but because it wants the same kind of control of this strategic material coming from Canada as it has when it is mined in the U.S. The U.S. has used the threat of its *Atomic Energy Act*'s requirement that only domestic uranium will be enriched to ensure that Canada is willing to export raw uranium or "yellowcake." The FTA therefore locked Canada into an explicit branch-plant relation to the U.S. nuclear industry.

Proponents of the FTA deal argued that Eldorado's (now Cameco's) uranium refining facilities in Ontario would be competitive. However, without the requirement of upgrading (i.e., refining) uranium exports, Saskatchewan's uranium yellowcake increasingly gets trucked straight south for refining and enriching in the U.S.

Continental, Not Environmental, Protectionism

It was hypocritical for the Mulroney Tories to accuse the opposition parties of using scare tactics during the FTA controversy. Throughout the debate the Tories used the threat of U.S. protectionism leading to the loss of Canadian jobs in the resource export sector as a major reason for negotiating the FTA. This scare tactic, which appealed to narrow, short-term self-interests, helped obscure how the FTA gave the U.S. assured access to Canada's "strategic" resources.

Until the U.S. developed its own uranium mines in the late 1960s almost all of Canadian-mined uranium went to the U.S. Since then the Canadian government and the uranium industry in Canada have tried to counter the U.S. legislation, regulation and litigation that protect the U.S. uranium industry. Since the expansion of uranium mining in northern Saskatchewan, the election of the Reagan administration in the 1980s and that country's military and nuclear buildup, there has been a marked shift in U.S. policy. On June 15, 1988, the U.S. Supreme Court unanimously ruled against their own uranium industry, which had filed for mandatory restrictions on uranium imported from Canada.

At that time uranium imports from Canada to the U.S. were valued at about $300 million (U.S.) yearly. With uranium prices dropping, the U.S. no longer had a viable uranium industry, and the number of domestic operating mines had dropped from 362 to only three between 1979 and 1986. Some market analysts claimed that the uranium price would have to be tripled for the U.S. uranium industry to stay viable, and this didn't even begin to account for the many hidden costs that result from long-term waste management, growing human cancers and social impacts. Even with uranium supply greatly outstripping demand, the U.S. government and nuclear industry continued to stockpile cheaper uranium from Canada, while letting their domestic industry die. In retrospect, it is credible to speculate that this was to ensure enough uranium for both commercial reactors and the massive nuclear arms buildup under Reagan's presidency.

Conflict in the U.S. over protectionism versus continentalism in the uranium industry was one of the major issues that caused Reagan to delay sending the FTA to the U.S. Congress. Though any tariffs on Canadian uranium would likely be overruled under the FTA, the use of U.S. uranium subsidies to maintain a domestic uranium industry continued to be hotly debated in Congress. The proposed U.S. Uranium Revitalization Fund provided $1.75 billion, mostly to fund uranium tailings reclamation work at closed mines. This spending wouldn't be a direct threat to the Canadian uranium industry and yet would appease some of the regions that had previously benefitted from U.S. uranium mines. It also showed how, even with the Cold War waning, the U.S. uranium industry had not even begun to cover the full environmental costs of its mining. There is a stark lesson here for Canadians: the meagre corporate taxes and revenues going to governments and the money set aside by industry for decommissioning mines do not take full account of the long-range costs of uranium mining.

The fact that a half century after the Uranium City mine in northern Saskatchewan was developed there was still no federal-provincial agreement on funding to "clean it up" didn't give the public much confidence regarding the handling of the mines after their closure.

Privatization and Foreign Investment

Notwithstanding the content of the deal, the FTA itself had a severe impact on continental economics. Privatization is a key component of the neo-liberal agenda and, in the case of Saskatchewan uranium mining, the connection between "free trade" and privatization is clear. The consolidation of the two Crown corporations, the SMDC and Eldorado Nuclear into Cameco, in 1988 and the subsequent privatization of this huge uranium firm through public share offerings harmonized perfectly with the agenda of those wanting to create an enlarged continental marketplace. It is no accident that the Institute for Saskatchewan Enterprise, which advocated privatization of the Saskatchewan Crowns in the 1980s, had managers from several resource export firms on its board — Sask Oil, the Canadian Petroleum Association, Saskatchewan Mining Association and the new uranium firm, Cameco, whose president was vice-chair of the Institute. The FTA and, after 1994, the North American Free Trade Agreement (NAFTA) served their interests. Government cuts, privatization and increased integration into the U.S. economy as an energy and resource supply state were all part of the new push for unfettered corporate enterprise and globalization. We see this corporate continental energy agenda gaining further ground under the Harper minority government, which took power in 2005.

The FTA was therefore much more than a mere trade deal. Even Canada's assistant negotiator at the time admitted the FTA went beyond trade, making even more hypocritical Mulroney's claim that the opposition was misrepresenting the deal. The "Overview" of the Agreement admits that the deal "should govern the trade and economic relationships [of Canada and the U.S.] for the foreseeable future." It acknowledges that, *along with the trade agreement*, the inclusion of services, business travel and investment is "ground breaking."

In particular, the section on "Investment" shaped the impact of the FTA. Under that section, "U.S. investors in Canada will be subject to the same rules as domestic investors when it comes to establishing a new business." While this was presented as being another "mutually beneficial framework," it reinforced Canada's role as a resource hinterland. "National treatment" for U.S. firms operating in Canada to acquire resources such as uranium (or oil and gas) for use in the U.S. gave them added control of our resources, economy and environment.

The FTA also included profound changes to Investment Canada, the review process that replaced the Foreign Investments Review Agency, which was abolished by the Mulroney government. According to the government's FTA *Explanatory Notes*, after 1992, direct acquisitions of Canadian businesses were reviewed only if the value of the gross assets was over $150 million. For indirect acquisitions, the threshold was raised to $500 million by 1991, and after that, according to the FTA, "there shall be no review."

Under Annex 1607.3, the uranium industry, along with gas and oil, was exempt from these changes, which meant a takeover by a U.S. uranium firm with assets below $150 million would still require a review. The existing practice of requiring that "a minimum level of equity be held by its nationals" would also still apply, and, under Article 1603, certain performance requirements of U.S. firms not allowed under the FTA would be allowed. These exemptions provided some special protections for the Canadian uranium industry, but even without enhancing direct U.S. ownership or control of the Canadian uranium industry, which remained possible under the FTA and now NAFTA, the U.S. gained guaranteed access to cheap uranium in Canada.

No Nuclear Non-Proliferation

Article 903 of the FTA prohibits "any tax, duty or charge on the export of any energy good to the other party, unless such tax, duty or charge is also maintained or introduced on such energy good when destined for domestic consumption." Furthermore, Article 904 allows:

> a restriction... with respect to the export of an energy good... only if... the restriction does not reduce the proportion of the total export shipments of a specific energy good made available to the other Party relative to the total supply of that good... as compared to the proportion prevailing in the most recent 36 month period....

This is tantamount to saying that Canada would have to continue to sell uranium to the U.S. in consistent proportions and at the existing market price. In other words, should Canada want to shut down the nuclear industry for reasons of global ecological protection or to seriously pursue nuclear disarmament, the FTA presented a major stumbling block to following such a path.

Articles 907 and 2003 of the FTA provided some national security exceptions to this export requirement. However, these clauses provide no basis for banning uranium exports to stop the proliferation of radioactive tailings, high-level wastes and weapons-grade material. Nor does reference to respecting the Non-Proliferation Treaty (NPT) provide much assurance regarding nuclear arms control or disarmament. Even with the Treaty in effect, and without the guaranteed access to Canadian uranium provided by the FTA, there is ample evidence, which will be reviewed in chapter 18, to show that Saskatchewan uranium has found its way into the U.S. and other nuclear weapons programs.

Around the time of the FTA controversy, the November 5, 1988, *Globe and Mail* reported the International Atomic Energy Agency (IAEA) acknowledging that Vietnam, North Korea, Columbia, Argentina, India, Pakistan, Israel and South Africa either already had sufficient material to make weapons or participated in "unsafe guarded activities capable of producing direct-use nuclear materials." Furthermore, in its 1987 *Annual Report* the IAEA admitted that it had verification problems "even with countries taking part in its inspection program."

The U.S. Navy uranium reserves for its submarines became exhausted in the 1990s, and with the shutdown of U.S. uranium mines, guaranteed access to Canadian uranium became a U.S. military necessity. Under a resurrected "Star Wars" the U.S could produce thousands of additional warheads with the help of Canadian uranium. And this is no hypothetical matter, as, according to the October 6, 2006, *Globe and Mail*, the Senate Committee on National Security recommended that Canada join with the U.S. "anti-missile shield." Even if Canadian uranium was not directly used in such a dangerous arms buildup, it would, as in the nuclear arms race of the 1980s, free up strategic material to do so. The FTA therefore not only consolidated economic continentalism, but it perpetuated our military and nuclear integration with the U.S.

In 1994 the North American Free Trade Agreement (NAFTA), which included Mexico, superceded the FTA. In Annex 602.3 of chapter six, Mexico reserved its right to develop its own nuclear fuel system separate from NAFTA. However, Annex 608.2 says:

> Canada and the U.S. shall act in accordance with the terms of Annexes 902.5 and 905.2 of the Canada–U.S. Free Trade Agreement, which are hereby incorporated into and made part of this agreement for such purposes.[2]

So, after NAFTA, regarding uranium, we remain a blatant branch-plant of the U.S. military industrial complex.

There has been one NAFTA dispute over uranium between Canada and the U.S., resulting from Russian uranium from dismantled nuclear weapons being "dumped" in the U.S. market. After reassurances from the U.S. in February 1995, Canada dropped its consultations and the issue never went to a panel. Clearly, those wanting to see a North American union grow out of "free trade" will want the continued transportation of uranium from Saskatchewan to the U.S. as a central part of the NAFTA energy superhighway.

Notes

1. See details in Jim Harding, "Jobs to Northern Indigenous People," *Aboriginal Rights and Government Wrongs*, pp. 26–30.
3. Available at <www.sice.oas.org/trade/nafta/chap-06.asp> accessed July 2007.

chapter 7

NORTHERN OPPOSITION MOUNTS

The Wollaston Lake Spill: A Calamity of Contradictions

When the Cluff Lake Board of Inquiry started its proceedings in 1977, Métis and First Nations groups unanimously called for a moratorium on uranium mining to allow for settlement of land claims, but because the NDP government excluded Aboriginal rights from the inquiry's mandate, these calls were dismissed. Left with the message, "take uranium mining or risk getting nothing," these groups shifted their focus to gaining assurances for employment.

With the memory of the Key Lake spills fading after 1984, the uranium controversy again slipped from public view. Meanwhile a rash of new uranium mines were being developed in northeastern Saskatchewan. The first was at Collins Bay, near the existing Rabbit Lake mine, a high-grade open-pit mine started in the 1970s. The Collins Bay mine was just off Wollaston Lake, one of the largest lakes in the province's north. Across the lake, the Hatchet Lake Chippewyan Band and Métis resided in the town of Wollaston Post.

The Collins Bay mine was of little concern to anyone other than committed non-nuclear activists. However, Greenpeace's 1988 campaign helped raise awareness of the proposed mine near Cigar Lake, which was just fifteen minutes by air from Wollaston Post. At the time this mine had the highest grade ore — averaging 12% — of any uranium mine in the world.

Once uranium mining faded from the limelight and the promised employment levels failed to occur, more disillusionment and resentment set in. Concerns deepened regarding the lack of benefits to northerners as well as environmental hazards from more uranium mines near Wollaston Lake. In June 1985 the Lac La Hache Band held a four-day blockade of the road to the Rabbit Lake and Collins Bay mines, and though the protest fragmented, the north wouldn't be muzzled for long.[1]

In 1989 another major uranium spill occurred, this time right at the Rabbit Lake mine. Like the Key Lake disaster in 1984, this spill received national attention. Mishandling of the public relations crisis exposed the failings of the regulatory system operated by the Atomic Energy Control Board, and even normally onside Saskatchewan daily newspapers began to question the viability of uranium mining. Up to this point several new mines in the vicinity had been proceeding with little public scrutiny. After the spill, many more groups began to call for a new more thorough uranium inquiry.

Northern Indigenous groups became more adamant about their concerns regarding the long-term impact of the uranium industry on their land base. The Wollaston spill had brought the uranium controversy back to life.

On November 6 and 7, 1989, 2 million litres of contaminated water spilled at the Rabbit Lake mine from a pipeline that stretched 10 km to the Collins Bay B open-pit mine, which was no longer being worked, to two settling ponds. The size of this spill was equal to 440,000 gallons, or 2,000 cubic metres — the capacity of three Olympic-sized swimming pools. The flow in the pipeline was about 3,700 litres per minute.[2]

During the month of November statements made by company, provincial and federal officials contradicted each other. It is unlikely the reasons for the spill will ever be fully understood without a public inquiry, though it appears that a cast iron valve located in one of the monitoring stations along the pipeline failed. There was either a failure in the actual instruments or a failure to correctly interpret the double instrumentation, which showed the pressure in the pipeline. Apparently an official on patrol passed the broken pipeline several times without detecting it. Simply put, this fail-safe system of "state of the art" mining failed.

When the spill occurred, the Rabbit Lake mine was not operating, having been shut down for modification and upgrading, and there had been a major layoff of mine workers. Nevertheless, Cameco continued to pump water from the open pit. Due to an inadequate monitoring system, the spill went undetected for sixteen hours. It overflowed the catchments designed to protect the surrounding land and went over the frozen ground to Collins Creek, about 300 metres from the pipeline, which flows into Collins Bay on Wollaston Lake.

Two days after the spill, Saskatchewan's Tory Environment Minister, Grant Hodgins, publicly stated that the levels of radium, arsenic and nickel in the spill were below provincial safety standards. Although there had not been time for a full investigation, he stated that the company was not guilty of willful neglect and would likely not be charged.[3] The next day the federal minister responsible for the regulation of uranium mining, Energy Mines and Resources Minister Jake Epp, stated in the House of Commons that he wanted "more questions answered on this issue," and he assured the House that there would be full disclosure to the public about the incident.[4] Later, however, Epp was to reject the call for a public inquiry made by the Hatchet Lake Band, District Chiefs and a broad array of provincial and national bodies. Minister Epp also said he'd been told the mine's monitoring system wasn't working. However, Cameco's public relations official, Rita Mirwald, was already contradicting this by publicly saying it was the result of an error in operating procedures.

The media was fairly quick to call for concerted action by the provincial government. In Saskatoon, where several uranium companies were headquartered, the *Star Phoenix,* in its November 14 editorial, "Record of spills causes concern," called for a tightening of conditions in Cameco's mining licence. This atypically tough-minded editorial also called for charges against the company, adding, "until corporate officers can be thrown in jail for environmental violations, fines may not prove enough of a deterrent…." The editorial noted that there had been several spills at Cameco's Rabbit Lake and Key Lake mines and that even after the large spill at Key Lake in 1984, there was an insufficient tailings storage area for expected waste production. The editorial ended with, "no matter how carefully government regulates uranium mining, it cannot prevent accidents. This leads to the obvious question of whether the benefits of uranium mining outweigh the risks." To my knowledge this was the first time the Saskatoon paper had speculated publicly that uranium mining might not be worth the trouble.

The *Star Phoenix* persisted with this unusually critical approach with its November 18 editorial, "Cameco should be charged for spill," calling for the provincial government to "lay charges against Cameco." Challenging many of Cameco's public statements, which tried to downplay the company's overall responsibility for the spill, the editorial stated, "There is a provincial process in place to deal with environmental violations and it should be used." It also hinted that a public inquiry "could well be the next step." The hard-hitting editorial ended: "If uranium companies can't handle the simple technology of a pipeline, how can the public be assured that they can safely deal with the highly-mechanized and computerized mining of the dangerous high-grade ores." Critics of the industry have consistently embraced such a common-sense, ecological approach, but it was unheard of in this usually pro-corporate newspaper.

Widened Call for New Inquiry

On November 15, several elected NDP officials called for a public inquiry into the spill at Wollaston Lake. (The Saskatchewan NDP was out of power at the time.) Ray Funk, NDP Member of Parliament for Prince Albert-Churchill River, was the first to make such a call. He stated that the uranium mining regulatory process had not been working and that the AECB had been aware of the inadequacies in Cameco's monitoring system for at least five years, adding, if that were true, "they're as much culpable as the operator."[5]

Funk also expressed concern about the proposed new mines at Collins Bay, Cigar Lake and Eagle Point being approved without any public hearings. The next day, Ed Benoanie, chief of the Hatchet Lake Band, called for a public inquiry to examine mine safety, federal regulations and the long-term impact of uranium mining on the northern environment. The Prince Albert District Chiefs and even the adamantly pro-uranium NDP MLA for

Cumberland, Keith Goulet, supported him. Chief Benoanie had previously given qualified support to the nearby mines but said northern bands were now strongly opposed to any further uranium mines unless they could be guaranteed no such incidents would happen again. He noted, "Ten years ago they said nothing is going to happen to the environment. That was ten years ago and today they say, oh these were bound to happen."[6] Ed Benoanie was replaced as chief in 1991, and there was some speculation about whether there was any influence by the very powerful uranium industry in the region.

There was a lot of bureaucratic stalling before Chief Benoanie was officially informed of the spill. Had he not been flying over the Collins Bay mine site the day of the spill and then made calls and inquiries, it would have taken much longer for the local people to find out what had happened. About a week after the spill, more information was released about the failure of the monitoring system and level of contamination. Despite Cameco's claims, AECB reports showed that Cameco's monitoring equipment hadn't been working when the leak occurred and for some time before that. Furthermore, the AECB's head of waste management confirmed that the leak had gone unnoticed while an inspection crew passed the location of the leak thirteen times in a sixteen-hour period. His conclusion: "What worries us significantly is some of the precautions we thought were supposed to be in place obviously weren't working properly."[7]

These revelations confirmed past allegations about company and regulatory irresponsibility. And they were announced by none other than Grant Hodgins, the same Tory minister who had stated a week earlier, prior to any thorough investigation of the spill, that the company was not likely to be charged due to there being no willful neglect. One would think failing to have monitoring equipment working would constitute willful neglect!

Even with both the provincial and federal regulators now saying publicly that Cameco was clearly at fault, Cameco's public relations spokesperson Rita Mirwald insisted that the mill site monitors were working. In spite of the obvious mix-ups that preceded the spill and continuing contradictory public statements by government and corporate officials, Saskatchewan's Environment Minister stated he was still not prepared to support the growing call for a public inquiry.

After these revelations about the breakdown of the monitoring and regulatory system, the Regina *Leader Post* added its editorial voice to the growing number of organizations calling for a public inquiry. Its November 17 editorial, "Public airing may clear uranium mine qualms," noted that it had been a decade since the last public scrutiny of the uranium industry and that a series of spills had occurred in the interval. The Regina editorial was more muted in its criticism of the uranium industry than its sister paper in

Saskatoon, saying only, "It may be as much to the benefit of the uranium industry… as to the critics of the industry to undertake another airing of issues relating to uranium mining in the province." The editorial simply repeated the view that the spill had "not resulted in unacceptable levels of radium, uranium, arsenic and nickel contamination." This defence of the status quo may have helped convince the two levels of government that it was not too risky to call for a public review of the new mines.

Debating Standards of Failing Regulations

The previous day a *Star Phoenix* story,[8] based on a statement by an official of Environment Canada, had reported that the contamination of radium had been ten times the provincial limit. Provincial samples showed the spill wastes contained 3 to 5 becquerels of radium per litre of water compared to the provincial standard of .37 becquerels per litre. (A becquerel is the new standard international unit replacing the curie and refers to the activity of a quantity of radioactive material in which one nucleus decays per second.) The official noted that federal measures, which looked only at dissolved radium, showed .24 becquerels per litre, well within the federal and provincial limit. The federal measures, however, are designed to monitor uranium mill effluent after it has been treated, which wasn't appropriate for measuring the untreated water in a spill. In contrast, the provincial measurement included radium particles. The Environment Canada official indicated that the chance of environmental damage was minuscule because the radium 226 would be greatly diluted in Collins Creek. (Radium 226, an alpha emitter with a half-life of 1600 years, is part of the uranium decay series.) He said the particles might add slightly to the natural radium on the lake bed, but would have no bearing on the quality of the drinking water in Wollaston Post, on the other side of the lake.

The public reporting of levels of radioactive contamination is usually an added source of confusion and anxiety in the aftermath of such spills. Some of the confusion comes from outright contradictions and denials, but some of it comes from different measuring methods and/or different standards. For example, it was announced that measures taken 600 metres downstream from where the spill entered Collins Creek showed a level of radium 226 of .11 becquerels per litre of water. The provincial standard for such a location would be .01 becquerels per litre. This was a smaller level than the .24 becquerels per litre reported by Environment Canada, whose sampling system allowed for more dilution of the contaminants. Even the smaller measure was still ten times higher than the provincial standard. There was some speculation that when the Saskatchewan Environment Minister initially (and falsely) said the spill was within provincial limits, he might not have understood that .11 is larger than .01.

There is some irony in the unfolding of this environmental dispute. Just prior to Chief Benoanie flying over Rabbit Lake, the AECB had apparently chartered a flight to undertake a previously scheduled inspection of the mine. The coincidence may have had a positive outcome. However, if the discovery of spills or other problems at Canada's uranium mines depends on sporadic visits by the regulator or a local chief, there is little hope for the surrounding areas. The AECB was clearly caught in a Catch-22 situation. According to the May 28, 1989, *Globe and Mail*, the AECB's submission to the Treasury Board indicated that it lacked the resources to adequately monitor and regulate the uranium industry. At the same time, during the public relations crisis caused by the spill at Wollaston Lake, the AECB had to present itself as a tough and reliable regulator for it and the industry to appear credible.

During the public controversy over the spill, the Saskatchewan NDP was holding its provincial convention. In government from 1971 to 1982, the NDP had spearheaded the massive expansion in the uranium industry in the north. While the NDP was in opposition, from 1982 to 1991, there was a steady decline in the popularity of the Grant Devine government, in part due to its lack of credibility in environmental protection, particularly over the handling of the Alameda-Rafferty Dam project.[9] The provincial Tory government's ignoring of the Federal Environmental Review's concerns regarding this dam project prompted the federal environmental panel to resign.

At the time of the Wollaston spill, the NDP was preparing to govern once again, and the controversy clearly created some consternation for the old guard in the party, who still harboured fantasies of co-owning a lucrative uranium industry. The NDP's 1983 turn-around of this policy had stressed no more uranium mines and the phasing out of existing mines while creating alternative employment. This policy had not been given much public profile during the 1986 provincial election; yet at the party's 1989 convention, one of the few points of contention was uranium mining. It's hard to say whether the Wollaston spill was the reason why there was so little support for a union-sponsored motion opposing the phase-out of existing mines. Even so, party leader Roy Romanow put a different spin on an adopted resolution that continued to call for no more mines and the phasing out of existing ones. He was reported to have said that the NDP does not endorse shutting down uranium mines without new jobs for the workers.[10] This interpretation would give a new NDP government not wanting to phase out existing uranium mines lots of latitude — including playing a passive role in the northern economy, with the outcome ensuring that alternative employment wouldn't exist.

The premier-in-waiting was also quoted as saying, "We mustn't judge whether or not it can be done in northern Saskatchewan by what the Tories have done because, by that standard, it can't be done." If Romanow meant uranium mines could be operated more safely under the NDP and that a phase-out would therefore not be needed, it contradicted existing NDP policy in place since 1983. In spite of the overwhelming opposition to uranium mining in the NDP rank-and-file, a major battle was still to be waged with the old guard, who would come to assume a lot of power in the new NDP government. The comment made by Romanow suggested it was likely that as premier, he, along with his inner cabinet, would work towards a policy reversal.

For Saskatchewan politicians, uranium mining is more a political football than a serious matter of ecological sustainability. A few days after the 1989 spill at Wollaston Lake, the NDP caucus's environment critic Ed Tchorzewski commented, "I think it's serious. I call it a massive spill of radioactive water." He also noted that, according to Saskatchewan Environment statistics, there were eighteen spills at uranium mines during 1988–89. What he did not say was that there were also spills, most notably the huge Key Lake spill, when the NDP was in power. Furthermore, it was the public outcry against an attempted NDP cover-up of a 1981 PCB spill in Regina that had forced the NDP government of the day to create a spill-reporting system. Only because of grassroots pressure have we become aware of the actual number and details of toxic spills.

Partisan politicians always blame "the other guy," while toxins from the industry they own and/or regulate accumulate and the planet's creatures and entire eco-systems continue to be threatened. But at least after the spill at Wollaston Lake, more northerners were beginning to realize that uranium mining was "hard ball" and that the game was occurring in their backyard.

Government-Industry Collusion Backfires at Baker Lake

Stringent opposition to uranium mining was soon to come from even further north. At the International Uranium Congress held in Saskatoon in June 1988, delegates from southern Canada and Western Europe heard from Canadian Inuit of plans by the West German company Urangesellschaft for a uranium mine near Baker Lake, North West Territories (N.W.T.), now a part of Nunavut. The other partner in this venture was Central Electric Generating Board Exploration, the mammoth British nuclear Crown corporation, with offices in Calgary. The Congress passed a resolution opposing this mine, and delegates from the N.W.T. and Saskatchewan and "Greens" from European uranium-importing countries agreed to coordinate their opposition.

The uranium industry and its government supporters were now not as free as they had been to disseminate pro-uranium disinformation in isolated northern areas, and alternative information typically withheld by proponents could now more easily be introduced into the local debate. The Keewatin Regional Intervention Coordinating Committee had already been created to try to make alternative information available to the sparse population of Inuit people dispersed across the huge eastern Arctic. Its membership included the Keewatin Regional Council (KRC), Inuit Association, Regional Health Board, Tungavik Federation of Nunavut and the local N.W.T. MLA. The Keewatin Wildlife Federation had already raised questions about the impacts of the proposed uranium mine, and soon the Tungavik Federation of Nunavut requested that the federal Department of Indian Affairs and Northern Development (DIAND) arrange for an information session on the proposed mine.

This was finally called for on March 1–2, 1989, in Baker Lake. People also came from Rankin Inlet, Eskimo Point and Chesterfield Inlet. Several territorial officials came from Yellowknife, the capital of the N.W.T. And several federal government officials, enough to bulge the small Igloo Hotel in the hamlet, flew in from Ottawa.

It was advertised throughout the eastern Arctic as an information workshop "to fulfill a commitment" to several Inuit organizations to provide background to help consider the uranium mine proposed by Urangesellschaft. The project, called Kiggavik (meaning "lone gull"), would be about 75 km northwest, upwind and up water from Baker Lake, the largest inland Inuit community in the Arctic, inland from Hudson's Bay. It would surprise most Canadians to learn that, though located 1500 km north of Winnipeg, Baker Lake is the geographic centre of Canada.

It didn't take long to see that the crass collusion of government and industry was going to backfire. The tone of the workshop was set when an official from the AECB, who headed up the uranium mine-mill waste management division, was seen at the back of the hall setting out promotional material from the CNA and AECL along with AECB material. There was obviously little concern about the nuclear industry (CNA, AECL) and government regulatory body (AECB) being so closely associated. When I asked the AECB official what this implied regarding collusion between industry and regulators, he simply replied that the pamphlets were actually brought by the Chamber of Mines, an industry group which meets with the Government of the N.W.T., which further showed the close operating connections between business and government.

Suppressing Translation and Controlling the Agenda

It proved naive to expect any balance or fair play, as no mining or government material was provided in the Indigenous Inuktitut language. Nor was the agenda broadly circulated or translated, and oral translation services didn't arrive until the evening of the first day of the two-day workshop. An official from the DIAND, a geologist who headed the Minerals and Economic Analysis program and supported the go-ahead of the uranium mine, controlled the agenda and meeting. The DIAND never negotiated the agenda with the Inuit organizations for which the meeting had, supposedly, been called. Exemplifying blatant paternalism, participants were told the government-run meeting was held "to inform them," that the DIAND official would chair the meeting and make concluding remarks, that the two main items for discussion would be "uranium and radioactivity" and "uranium mining operations." The only resource persons listed on the agenda were from government and industry.

However, at the request of the Keewatin Regional Council, the DIAND paid for two other resource persons to be brought up. As Saskatchewan uranium mines were, apparently, to be used as case studies by the proponents, I was asked to come to discuss Saskatchewan's public inquiries, socio-economic impacts, environmental pollution and uranium end uses, subjects I had researched for a decade. Upon arriving I was asked if I had a business card to be copied along with those of other resource persons, but I then found out there was no place for me on the agenda, which seemed a strange way to encourage balanced public participation or to provide a context to allow broad informed discussion of the proposed mine.

When I asked the chair when I might present, I was told "tomorrow, during 'end use.'" The next day the KRC was told they'd be able to have their resource persons present after lunch; later they were told after supper. Time kept slipping away. At 9:00 p.m. on the second day, the DIAND chair adjourned the meeting before the KRC had the chance to inform the audience that there were other resources present and available. The DIAND clearly did not want the resource persons brought in by the KRC to be associated with the "government-run portion," and, as it turned out, the federal government-run portion was the only portion allowed.

More Propaganda on Radiation

The AECB began its presentation on "uranium and radioactivity" with a ten-year-old U.S.-made video on radiation and nuclear power. It downplayed the dangers of radiation by repeating that radiation was "natural." The message was pounded home: "It's all natural and found everywhere on earth"; there's been a "great natural flow of radiation going on since the beginning of time"; "we get radiation naturally"; "it's been here since the earth was formed"; radiation exists "naturally all over the universe."

With this barrage one could easily forget that stable biological evolution didn't occur on earth until the background radiation levels were greatly reduced or that the integrity of biological life still depends on protection against radiation from outer space and from the deep earth. The destruction of the ozone layer is already allowing in more ultraviolet radiation, with consequent significant increases in skin cancer. Mining and milling of radioactive rock is bringing more radiation into the biosphere, with significant increases in lung and other forms of cancer. These are all natural phenomena; water, too, is natural, but when it fills our lungs, we drown.

The video failed to mention that military and industrial activity during "the atomic age" had already added to the so-called natural background levels through such things as atomic testing, mining and milling of uranium, the buildup of nuclear wastes and reactor emissions and leaks. It was blatantly one-sided on the ongoing scientific controversy over radiation and public health. White-coated experts featured in the film asserted, emphatically, that "Very large amounts of radiation can be dangerous" but that "as radiation goes down, the risks go down." Much was made of the role of radiation in cancer therapy, but nothing was made of the incidence of childhood leukemia from past permissible levels of x-rays; nothing was said about higher levels of lung cancers among uranium miners; nothing was said about the fact that permissible radiation levels have had to be reduced ten-fold since the 1900s, and that the United Kingdom had recently reduced permissible levels to one-third of its previous level, or Canada's existing level.

The video was designed to make people feel comfortable with the radioactivity of the nuclear fuel system, as a prelude to accepting the proposed uranium mine. Like most nuclear propaganda, it was based on withholding information. So I challenged the AECB's "health physicist" about the statement in the video that "as radiation goes down, risks go down" and the video's failure to mention the special risks of lower levels of alpha radiation, especially when inhaled. She simply defended the U.S. video.

I then asked her if she had seen a recent letter from Karl Z. Morgan, the founder of health physics, which pointed out that Canada's Minister of Energy, Mines and Resources, Marcel Masse, was "badly misinformed on the matter of risks from human exposure to low levels of ionizing radiation." She asked who Dr. Morgan was. (He was a founding member of the International Commission for Radiation Protection, which sets the standards that the AECB uses and defends.) When I asked why she hadn't discussed "ionization" in her talk on radiation, she commented that she didn't think the local people would understand. I asked her what she expected without translation services. I then quoted a portion of Dr. Morgan's letter: "There are more radiation-induced cancers per unit dose… at low doses than at high doses…. All mixed models tested did much better than the linear models." (The linear model suggests the risks are proportionate to the dose, whereas the mixed models test assumptions about disproportionate risks at lower levels.) Her response was that she could find scientists who would say the opposite. I asked her "who?" She was silent.

Later I went back to this AECB regulatory official to ensure that what she was saying was that risks went down proportionate to radiation dose, often referred to as the linear hypothesis. She said she was. I then read to her from one of the pamphlets in the AECL's Nuclear Information Series placed on the table at the back of the meeting, purportedly to help inform the Inuit. That statement, from the pamphlet, "Radiation is Part of Your Life," said: "For alpha radiation the linear hypothesis is less likely to overestimate the risk and, in fact, may underestimate it."

Though this publication promoted an industry perspective, it seems that by the mid-1980s even some nuclear industry-employed scientists had to acknowledge the new research on the greater relative risks from lower levels of radiation. The question remains how a person employed by the federal government as a "health physicist" could be so unaware of the underlying scientific and policy issues. Perhaps the lines had become totally blurred between regulation and public relations.

The president of the Keewatin Inuit Association, Louis Pilakapsi, also criticized the AECB for its use of this U.S. video. When the AECB official was further challenged about the overall pro-nuclear bias of the video, he finally "went on record," as he put it, that "some of it is propaganda." This was his, not anyone else's, term. The question remains why a publicly funded regulatory body — whom the public would assume kept itself at arm's length — would see itself as a promoter of the nuclear industry. This encounter with the AECB put new meaning into their motto: "To ensure that Canadians can benefit from nuclear energy without undue exposure."

After a morning of skirmishes over the AECB not being very candid about radiation risks, the chair began the afternoon session expressing concern about being "sidetracked by questions." He wished to move on to what he called "more pertinent questions," forgetting that the morning agenda on radiation and uranium mining was his agenda, with his resources. Things obviously hadn't gone the way he wanted.

Moose Aren't Caribou

After defensively declaring he was "not here to advocate for the mine," the chair stressed that the meetings should move on to look at the actual plans for the mine. Before doing so, however, he again called on the AECB official, this time "to inform us" about air pollution. The AECB argued that water was the major environmental pathway for uranium mine pollution and admitted that ore from mining, milling and waste tailings were all a "source for water and air contamination." He said both dust and radon gas were sources of air pollution and suggested the risk from radon gas was more of a "problem for workers than for the environment."

To prove his point he referred to a study of radionuclides in moose organs in the Elliot Lake, Ontario, area. He referred to this as a worst-case scenario, since there had been twenty years of weak regulation of uranium mining, and there were already 160 million tonnes of uranium tailings on the surface. This study apparently found "no statistically significant differences" in radiation levels in moose killed around Elliot Lake or elsewhere in the region. The radiation levels weren't given, nor was it admitted that moose travel widely, could not easily be assumed to be "Elliot Lake moose" and, as such, the effect of air-born radioactivity in the food chain could be averaged out.

Many Inuit people persisted in asking why the AECB was concentrating on moose and not addressing the risks to the caribou herds that populate the Keewatin region. One hunter reminded the AECB official that there is no bush in the Keewatin tundra and that the diet of the caribou is very different than moose, with Caribou depending heavily on lichen, which bio-accumulate radioactive caesium. He reminded the meeting how quickly the radioactivity from atmospheric atomic tests in the 1950s and 1960s had shown up in the Arctic caribou, upon which most Keewatin Inuit depend for food. Furthermore, it was mentioned that the air pathways to caribou and the Inuit food chain weren't only from mines in the region. DDT, PCB and lead concentrations in the Arctic show the northern hemisphere is particularly vulnerable to pollution carried in global weather patterns.

Throughout the discussion the AECB avoided any mention of Chernobyl or that 90,000 caribou were slaughtered in Sápmi (formerly Lapland) after the Chernobyl nuclear reactor accident due to high concentrations of radioactive caesium in their bodies. Sápmi is about as far north of Chernobyl as the Keewatin region is from the concentration of nuclear reactors in the northeastern U.S. and Ontario's industrial belt.

After these questions and comments on the global impact of uranium mining, the AECB official responded, "Lead is mined for various reasons, like batteries, gas and bullets. Does that mean we have to stop mining lead?" The question was obviously rhetorical. The implied analogy was between nuclear weapons containing uranium and bullets containing lead. In actuality lots of substances besides lead (including depleted uranium) can be used in bullets (or batteries), but only uranium or its fission-products like plutonium are used as fuel for reactors or nuclear weapons. Furthermore, lead in gasoline is itself a serious environmental and health problem, especially for young children, and is prohibited for this reason. This reduces demand for mining lead, as alternative energy and nuclear disarmament would ultimately shrink the uranium mining industry — baby steps towards ecological sustainability, which we are obliged to take.

This government-sponsored workshop, like the Federal Environmental Assessment Review Office (FEARO) panel that soon followed, ruled out discussion of the nuclear fuel system, including dangers from the end uses of uranium in reactors and weapons — omissions that clearly biased the discussion in favour of the uranium industry.

Earlier, a federal fisheries researcher doing experimental work on natural pathways of radionuclides discussed issues of water contamination from uranium. Water contamination from uranium mines, *per se*, hadn't been discussed. Uranium mine pollution was somehow being limited to a discussion of air pathways. But Tagak Curley, president of the Keewatin Wildlife Federation (and spokesperson for the Keewatin Regional Uranium Intervention Coordinating Committee) reminded the AECB about the destruction of the Serpent River system, downstream from uranium tailings at Elliot Lake, Ontario. The AECB official admitted these mines weren't properly regulated, but argued that acid production, not radioactivity, had killed the fish for sixty or so miles downstream from the mines. A participant wondered if the acid got the fish first, and were they still alive today, whether they show severe concentrations of radioactivity.

The discussion shifted to long-term problems from uranium tailings. The chair said Saskatchewan officials would explain this to us later. Industrial proponents of uranium mining in Saskatchewan in the 1970s sometimes argued that it makes little difference whether the natural radioactivity is in the ore or in the tailings. The AECB official, however, admitted "radiation is more available to the environment after mining." He also admitted that radioactive wastes are "active for a very, very long time." However, even after all the admission of problems of mine regulation at Elliot Lake, he assured the Inuit that it was the regulator's responsibility to make sure there was "no impact." It was that simple. Later he admitted there were only four AECB officials nation-wide dealing with waste management.

Radon gas and its short half-life (3.5 days) were discussed, almost as though its danger disappeared in a short time. There was no mention that it was possible for radon gas to travel 1,000 km with a wind of 10 km/hour before one-half of it had decayed. (The winds are much stronger than 10km/hr around Baker Lake). Nor was it mentioned that this radon gas was constantly being replenished from the radium in the tailings (with a half-life of 1,600 years), nor that the radium was constantly being replenished from the decay of thorium (with a half-life of 76,000 years). The uranium decay series was never discussed or explained directly, but only referred to in a few snippets, mostly in reaction to probing questions. No context was ever created for such a vital discussion.

Corporate-Style "Adult Education"

After an afternoon break, the chair called back the executive vice-president of Urangesellschaft to describe the project; from that point on, the company representative had more time at the front than all of the other resource people combined. However, if this was a temptation to railroad the meeting, it backfired. There still was no translation equipment, and several local people called on the chair to be more sensitive to the Inuktitut-speaking elders who had come to the meetings but were being left out of the discussion. The chair responded, "I'd hate to delay" the workshop, after which an Inuit man spoke up, "We've been waiting all our lives for things from government, a few hours won't make much difference."

The mining executive opened his talk by saying he was here "to give you information which we all know you need." This view, that learning is a one-way street, dominated his presentation. In old-school engineering style, he described the project in terms of its components, with inputs and outputs, as though it had no particular ecological home or impact, and certainly no connection to the global problems of the nuclear reactor and weapons industry. Nor, apparently, did the mine have any political or economic context.

It was all made to sound so straightforward. Barges would bring the massive tonnage of construction materials from Montreal, through Hudson's Bay and up to a marine terminal constructed six to eight miles east of Baker Lake. A winter road would move materials to the ore body. An on-site airstrip would fly in workers and food from Winnipeg. These workers would dig two deep open pits and take the higher-grade ore to a mill to remove the uranium. Sulphur would be shipped in to make sulphuric acid on site, for the leaching and neutralization process, and a limestone mine would be created, with lime brought to the ore site by another winter road. All of this was described in fail-safe terms, in stark contradiction to the common sense that filled the room of active hunters and trappers. No mention was made of accidents or spills until the audience raised such inevitabilities. The mine officials didn't even know that a major calving area for the caribou was where the limestone mine was planned.

It sounded credible, though upon careful scrutiny it was circular, technological rationality, which is how the exploitation and plundering of hinterlands throughout the globe has been rationalized. This "description" was, of course, all done with English overhead projections, which not even the English-speaking members of the audience could clearly see.

The chair then moved to the topic of regulating the mine and he called back as expert the AECB official, whose credibility was already eroded from the morning session. I began to realize that the DIAND and the uranium consortium probably thought that if the non-industrialized local Inuit people had uranium mining technology and government regulations explained to them in simple terms, their questions and opposition would somehow dissipate. The depth of their conceit and paternalism was unbelievable.

The AECB official referred to his work with the government regulatory body in "business" terms; at one point saying, "I've been in this industry for twelve years." Again, none of his material was translated into Inuktitut. Eldor's Rabbit Lake uranium mine in northeastern Saskatchewan was mentioned as the example of ideal regulation. Prior to it closing, this mine was often used as an example to argue that uranium-mining technology is fairly straightforward and can be left to the experts in industry and government. This non-unionized mine, with a commuting system for bringing in labour from the south, was a model for other uranium mines such as Cluff Lake and Key Lake. The seven-day-in seven-day-out work structure made it function as an ideal compliant labour camp, with no adjacent company town to develop a local culture and collective consciousness.

The AECB representative wanted to assure Inuit people that it would "regulate the same way [in the N.W.T.] as other mines." The questions and pressure from the audience, however, continued to mount. What would the AECB do about waste management, if the mining company abandoned the mine? After all, the spot price of uranium at the time of the Baker Lake workshop was only $11.60 a pound. This was about 25% of what it was a decade before, during the so-called uranium boom following the "energy crisis" created by the Organization of Petroleum Exporting Nations' (OPEC) huge oil price increases in 1973. Someone specifically asked what the AECB would do if the companies involved declared bankruptcy and abandoned the mine. The AECB responded, "It's never happened in fifteen years," and the mining executive intervened, "It would be unthinkable that we would run away from our responsibilities." Of course, even if bankruptcy and abandonment never occurred, little or nothing is known about safe, long-term uranium mine waste management, and as the company executive tried explaining the shifting waste management plans, participants in the workshop became even more skeptical.

Uranium companies like to emphasize short-term economic benefits in the hope that people who are facing desperation or motivated by self-interest will turn a blind eye to the longer-term ecological legacy. This strategy backfired at this workshop. In their project description, Urangesellschaft Canada Ltd. said it would "dispose" of the radioactive tailings between two drumlins. (A drumlin is an elongated hill or ridge of glacial drift.) Later they stated they would store the tailings in a lake and later "dispose" of them in the two open pits left from the mining. At the workshop the plan changed again — this time to construct containment berms with the waste rock in the hope that the permafrost would move up and stabilize the radioactive tailings.

After this session, even some the government proponents were privately expressing skepticism about the company's plans, but there was never an opportunity to question how the company would decommission the mine. Remember, 85% of the radioactivity in the ore remains in tailings. Discussion of the serious problems that occur after a profitable mining venture is over in ten years or so evaporated from the agenda, in much the same way the proponents wanted northern Inuit people to think the radon gas from the tailings would "evaporate."

After this set of exchanges, you could feel a distinct shift in the room. The president of the Keewatin Inuit Association, Louis Pilakapsi, intervened and commented that it sounded like the company was "bound and determined to proceed even if there was opposition." He said it seemed the "mine was in preparation already." He asked why mining was started when Inuit people had not okayed this or even been involved, raising the matter of impending Inuit land claims. For the first time in the workshop, he raised matters of social impact and noted that at Rankin Inlet there was a negative impact on the local Inuit community from the mainly white labour force at the gold mine. The town had become a government settlement after the mining company left, and he stated outright: "Inuit young won't benefit!"

He ended his statement by saying, "I'm going to be against uranium mining. We have to think of the future of young people." The thunderous applause, the first one of the day, was confirmation that the DIAND-run workshop was not going as planned.

At this point the DIAND-employed chair spoke on behalf of the mining company, explaining why it had gone ahead with site preparations. He commented: "It might not go ahead.... FEARO will decide.... There won't be any more Rankin Inlets." Actually, the FEARO only recommends to the federal Environment Minister, who then goes to Cabinet for a decision. It wasn't very convincing.

Finally, the mining executive, now more animated, intervened, saying, "I'm concerned that you suggested our company wasn't concerned about people in the North.... Life is more certainly more important than money. That is certainly our philosophy." After a pause, seemingly to ensure he had good control of his words, he continued, "I must say that it is purely a misconception regarding the suggestion that we are going ahead in spite of the feelings of local people." Then, after another pause, came the punch line: "In the final analysis, when it comes to the decision process, the decision to go ahead will only take place if we have the support of the local people."

This ambiguous guarantee didn't reassure the Inuit present. After all, the FEARO process, like the uranium inquiries in Saskatchewan, had already ruled out discussion of Inuit land claims. Another Inuit spokesperson asked: "Urangesellschaft has been here for fourteen years, why [is the mine] just coming out now?" He continued, in reference to the lack of adequate translation, "The elders don't understand.... I feel sorry for the elders here." Sparking another general round of applause, he said, "I am disappointed we are just starting to talk about it."

At this point the DIAND chair admitted, "You're right, it is a bit short." The Keewatin Regional Council executive officer then asked the mining executive for financial assistance to help get more information translated on the project. His response was: "I'm not sure that would be the best way to get information of a neutral nature on the project." He then referred to the need for information to be provided by "someone who is unbiased," by which he likely meant pro-nuclear government officials.

By this point the lines were pretty much drawn, and it was becoming a lonely and somewhat unsettling affair for the proponents. When we returned to the Igloo Hotel for supper, I overheard the mining executive saying, "It doesn't look good," talking long distance on the only pay phone in the cramped lobby. After a supper break and a bitterly cold walk back through a blizzard to the school auditorium (which reminded all participants of just how strong the wind was from the direction of the proposed uranium mine), the workshop reconvened. The company executive was immediately asked if he thought such a mine would ever be considered fifty miles from Toronto. Not unexpectedly the company spokesperson sidestepped the issue, simply saying that too would be settled by the government's FEARO process.

This debate over where the real decision-making power existed led right into presentations by Saskatchewan Environment officials. And though they still tried to paint a rosy picture of the safety and benefits of uranium mining, their so-called expertise was starting to fall on deaf ears. The Inuit people had already figured out, to their satisfaction, who and what to believe. Several months later, after continuing to express their skepticism in the federal public hearings, the people of Baker Lake voted 90% against the proposed uranium mine. This was just too much for the company, which had uranium holdings elsewhere, and the momentum for the mine project evaporated. Facing this level of resistance, the consortium looked for "safer" places for their investors. Like the Mennonites who had stopped a uranium refinery in Saskatchewan in 1980, the Inuit at Baker Lake had shown it is possible to stop the nuclear industry from expanding on their doorstep.

Notes

1. See Miles Goldstick, *Voices from Wollaston Lake*, Earth Embassy and World Information Service on Energy — WISE, 1987, Amsterdam.
2. "Rabbit Lake Spill Causes Great Concern," *The Northerner*, Vol. 35, No. 40, November 15, 1989, pp. 1–2.
3. "Uranium mine spill called not dangerous," Regina *Leader Post*, November 9, 1990, p. D10.
4. "Epp wants more answers on spill in Saskatchewan," *Globe and Mail*, November 10, 1989, pp. A1, 2.
5. "Mine spill prompts Funk to demand public hearings," Saskatoon *Star Phoenix*, November 15, 1989, p. A12.

6. "Inquiry into uranium mining urged," Saskatoon *Star Phoenix*, November 16, 1989, p. A12.
7. Randy Burton. "Cameco irresponsible in radioactive leak: Hodgins," Saskatoon *Star Phoenix*, November 16, 1989, p. A1.
8. Randy Burton, "Spilled water exceeded allowable radium levels," Saskatoon *Star Phoenix*, November 16, 1989, p. A12.
9. This pet project of Grant Devine involved damming the Souris River to create the Rafferty Reservoir and the Moore Mountain Creek to create the Alameda Reservoir. Built in south-eastern Saskatchewan between 1988 and 1995, it provides water for the Shand (coal) power plant and provides water and flood protection for the region.
10. Mark Wyatt, "Most policies spark little or no debate," Regina *Leader Post*, November 28, 1989, p. A4.

chapter 8

NFB RELEASES *URANIUM*

In 1990 the National Film Board (NFB) released its film *Uranium*. Completing the film was an uphill battle. Had it not been produced in Edmonton, Alberta, far away from the eastern head office and federal bureaucratic politics, it might never have made it to the screen. Its northern release turned out to be an intense political event. In October 1990 the film premiered in Yellowknife, Baker Lake and Rankin Inlet, N.W.T., then later at Wollaston Lake, Saskatchewan, where there was great interest due to the nearby cluster of uranium mines. It was also shown in Edmonton and Saskatoon. After each showing, a panel discussed the making of the film and the issues it raised. A detailed, well-referenced booklet, "*Uranium: A Discussion Guide*," was produced by the NFB in 1991.

At all these well-attended meetings, federal government and industry employees strenuously tried to discredit the film. I was a consultant in the making of the film and was asked to accompany the NFB producer Dale Phillips and director Magnus Isacsson to sit alongside them on these panels on the release tour. After travelling to several locations, I began to realize that the "attacks" were a coordinated and planned response. In Rankin Inlet, I was shown copies of faxes sent between federal bureaucrats, from Ottawa to the north, preparing for the release of the film. Energy, Mines and Resources (EMR), the AECB, the AECL, the DIAND and an array of uranium mining companies were all upset by the NFB's new release. Their concerns were about the film's handling of the issues of radiation, environmental impact, nuclear proliferation and Aboriginal rights. I kept detailed notes of their criticisms and later wrote and circulated a rebuttal, which is the basis of this chapter. In 1990 *Uranium* won the award for best documentary at Saskatchewan's Yorkton Film Festival.

Conflict of Interest Downplays Dangers to Northerners

The EMR had a big stake in the northern response to this film. Its ministry housed the nuclear regulatory body, the AECB, and had direct ties to the federal government's nuclear Crown corporation, the AECL. Minister Jake Epp, responsible for the nuclear regulatory system, had recently reversed the decision to phase-out the huge subsidies to the AECL, instead raising annual public subsidies to $220 million per year. Epp also had a political stake in the industry, as the Manitoba riding he held included the AECL's Whiteshell

Nuclear Station, which was vulnerable due to the lack of demand for nuclear power and was finally shut down in 1998. The pro-nuclear EMR therefore wanted more direct control of the final version of this film, and its senior bureaucrats were very disturbed that this arms-length, federal agency, the NFB, was so unwilling to toe the federal line on energy policy.

The film is an advocacy and educational film, creating balance in the public domain by presenting first-person accounts of people who have lived near uranium mines and tailings over the decades. Trying to avoid the appearance of being colonialist, federal agencies indirectly attacked Indigenous and Inuit concerns about uranium mining, suggesting that the film should have refuted statements and information the EMR deemed questionable. The Indigenous people interviewed for the film apparently were to be allowed to describe their experience at the front-end of the nuclear system but should then have been "corrected" by the pro-nuclear professionals.

In view of the highly slanted claims made in the $20 million pro-nuclear CNA promotions campaign initiated in 1987–88, it was hypocritical of the EMR, AECB and AECL to accuse the NFB film of being biased. Any objective survey of pro-nuclear and anti-nuclear information materials would plainly expose errors of commission, and most importantly of omission, of vital information relevant to the public good, on the pro-nuclear side. For every scientist working close to the industry who makes factual claims about the safety of the uranium industry, there is a scientist not working in the industry who can establish other facts that do not corroborate this view. While scientific truth-seeking takes many unexpected turns, the greatest ideological bias operates among employees who depend upon the nuclear industry's continuation.

In the uranium inquiries in Saskatchewan in the late 1970s, the nuclear industry and independent scientists faced off over risks from low-level radiation and the adequacy of existing radiation protection standards. The commissioners of the Cluff Lake and Key Lake inquiries, who were all inclined to be pro-nuclear, believed the industry scientists. However, even by the late 1950s, several radiation standard-setting bodies, including the BEIR (Biological Effects of Ionizing Radiation) and the ICRP (International Commission on Radiological Protection) had concluded the risks were much greater than previously thought.

But the EMR, AECB, AECL and DIAND weren't that interested in scientific or, for that matter, ecological factors. Rather their goal was to neutralize the negative effects the film might have in the north, but the ploy didn't work. The public in the north as well as in the south was becoming more aware of the limits of narrow, short-sighted scientism, which tends to promote costly mega-projects rather than sustainable, ecologically friendly alternatives.

The AECL probably had the most to lose from the film. It criticized the

NFB film for not putting uranium mining and the radioactive tailings left within the biosphere into the context of background radiation. Because background radiation is normal — or so the nuclear logic goes — there shouldn't be such a fuss made about "a little" additional radiation from the nuclear fuel system. At one of the panel discussions after the film's showing, a pro-nuclear scientist actually argued that if radiation didn't exist there wouldn't have been human evolution. Was this strange remark supposed to make radiation seem more benign to us? The nuclear physicist speaking with such an evolutionary sweep claimed strict neutrality from the uranium and nuclear industry, but immediately went on to paraphrase the industry view that nuclear energy is the solution to global warming. He was not knowledgeable about research outside his field of nuclear physics, such as that done by the U.S.-based Union of Concerned Scientists, which shows the impracticality and ineffectiveness of this strategy. According to his "naturalistic" logic, he might have concluded that since the water in the oceans is natural, why worry that rising ocean levels will flood coastal settlements and islands around the globe.

The biological vulnerability of humans to radiation, including background radiation, has consistently been underestimated by the industry. It has its own economic interest in keeping legal exposure limits from being set too low and thus made too costly. Remember that 35 rems per year was once a legal dose of radiation. In 1958 it was then brought down seven-fold, to 5 rems, on the basis of several studies. New research suggests a five-fold decrease is now required, to 1 rem (or 10 mSv using the new measure), which has not occurred in Canada to date.[1] That would constitute a thirty-five-fold decrease since World War II, a trend that doesn't give the public much confidence in continuing industry assurances.

The industry consistently claims that the conditions that led to excessive lung cancers among uranium miners no longer exist. Nearly 300 Ontario uranium miners had died of lung cancer at the time this NFB film was released and the number is still rising. It also claims that it can now operate below the existing legal level of radiation exposure. There's no disputing that exposure to deadly radon gas is lower in open pit mines in northern Saskatchewan than it was in the deep shaft mines of Ontario or in Saskatchewan's Uranium City. However, if it is the case that industry can operate below legal limits, why has there been so much resistance among both industry and nuclear regulatory bodies to reducing the legal limit? The AECL and AECB were aware of calls for a further reduction in radiation exposure units and of studies showing six to eight times the risk of cancer from exposure within existing limits. Why were they so close-minded on this topic at the same time as they accused the NFB film of misinformation?

Might these nuclear bodies want to keep enough flexibility, even if it presents serious risk to workers and the public, to ensure that they can mine the very high level, more profitable ore bodies, some still underground, such as at Cigar Lake in northern Saskatchewan? Imposing new, safer limits could make the industry even more uneconomic, and that, not human or environmental health, remains their bottom line.

The AECL was also concerned that the NFB didn't mention studies showing minimal environmental damage from uranium mining. The one and, to my knowledge, the only such study, which the industry referred to at nearly every hearing or meeting across the country, was done by Beak Consultants and showed no seepage of radioactive material into Wollaston Lake from the nearby Rabbit Lake uranium mine.[2] However, this study only covered ten years, hardly a basis from which to understand or predict the movement of radionuclides through eco-systems around the uranium tailings over several thousands of years. For another thing, the mine was operating during this ten-year period, and therefore there was some form of monitoring. Even with this monitoring, there was a 2 million litre spill of radioactive water at Wollaston Lake in November 1989.

Indigenous people and environmentalists are concerned about what will happen over many generations. Under present regulations, the abandoned mines only have to be monitored for five to ten years after closure, and there are no funds being put aside for long-term tailings management. Allan Blakeney's NDP government tried to have some foresight about this by creating the Heritage Fund, which was to build up from uranium and resource royalties, but this was depleted during the province's fiscal crisis in the 1980s.

It is telling that it has taken two levels of government over fifty years to come to an agreement on interim funding for some clean-up of tailings at the Uranium City mines, and no thorough clean-up has yet taken place at the Port Radium mine in the N.W.T. If the real long-term costs of tailings management or permanent control (if such a system were ever possible) were calculated into the price of uranium (which should be done in a sustainable development plan), the industry would immediately become uneconomical. But the AECL didn't want to talk about this.

The AECL was also distressed that the NFB film raises the historical and contemporary connections of Canadian uranium mining with nuclear weapons. The industry tries to disassociate itself from the proliferation of nuclear arms by claiming that uranium mined for fuel for the Candu reactor is never enriched, so it can't be diverted for weapons. Yet the major way the Candu has contributed to nuclear weapons' production is directly through the plutonium in its spent fuel, which can be accessed without a reactor shutdown. Furthermore, Canada's Candus now produce large amounts of tritium for export. Tritium is used in nuclear weapons, and the amount produced and exported from Ontario's Candus is far beyond the U.S. demand for civilian needs, so clearly the weapons connection remains problematic.

The industry's NFB film critics address the danger of proliferation of nuclear weapons from uranium exported for enrichment by hiding behind "Canadian safeguards" under the Non-Proliferation Treaty (NPT), which bans the diversion of uranium into nuclear arms. Their claim is that if Canada were to stop exporting uranium, the impact would be on reactors, not on weapons. There are several shortcomings to this view. Even if one could be totally assured that Canadian uranium never gets diverted after enrichment for use as fissionable material in nuclear warheads, the depleted uranium left after enrichment is commonly stockpiled and is available for use in military reactors, the casing of H-bombs and producing depleted uranium bullets, which, once used, permanently contaminate war zones with radioactive dust. I discuss this at length in chapter 18. Furthermore, there is the practice of swapping labels on shipments of uranium to meet present demand, making control of the end-use of Canadian uranium even more difficult. An analogy is the well-documented diversion of legal pharmaceutical drugs into the even more lucrative illicit market. In the case of Canadian uranium exports to France, where the commercial and weapons' production system are highly integrated, it is even more difficult to accept assurances that Canadian uranium doesn't find its way into weapons.

Because of the difficulties in guaranteeing that Canadian uranium exports aren't still being diverted for weapons, the industry often argues that there is a balancing of quantities of uranium, so there can never be more uranium available for weapons than that which comes from the available supplies from the weapons-producing countries themselves. The more sophisticated critics of the NFB film relied on this reasoning, but the argument is not convincing. The availability of Canadian uranium supplies to these countries for commercial purposes is what allows them to use their own supplies for weapons' production. Thus, Canada remains open to the charge of complicity in nuclear arms proliferation.

No uranium companies ever admit they produce for nuclear weapons, but the weapons-grade material must come from somewhere. Until there is complete nuclear disarmament, Canadian uranium explicitly exported, or diverted, for military purposes will remain in existing stockpiles of nuclear weapons. The industry just can't wash its hands of this issue.

Federal Government Paternalism Alive and Well in North

When the AECL addressed Aboriginal rights, which the NFB film implicitly highlights, some very fuzzy thinking ensued. The merits of the film were questioned because, in the opinion of the critics, Indigenous people faced "real" social and economic problems — as though the dangers presented by the expansion of uranium mining were not real. By the time of the film's release, over 2,000 jobs were supposed to exist from the new Saskatchewan mines and northerners were to get 50% of these under a surface lease agreement. Only about half of the promised jobs were created, in part due to the shutdown and layoffs at the Uranium City mines. Of these, about 30% overall went to northerners, and "northerners" continued to be interpreted broadly so as not to have to give preference to Indigenous people. My research found that only 200 of the 1,000 jobs in the operating mines by 1990 were held by northern Indigenous people.[3]

Also, great revenues were promised to the province to aid with northern development, whereas the actual revenue when the NFB film was under attack was only 6% of the highest and 15% of the lowest estimates made when the new mines were originally opened up. According to Saskatchewan's *Public Accounts*, in 1989, uranium revenue was $30 million instead of the $185–432 million predicted during the CLBI. And this pittance went into the government's slush fund; it was not earmarked for the north. The promise of jobs, revenue and other benefits back in 1978 appears now to be a cynical ploy to neutralize the call for a moratorium and opposition to the mine until companies could get government approval to get the mines on-stream.

The response of the DIAND to the depiction of Indigenous concerns in the NFB film showed that paternalism was not dead in the federal civil service. DIAND officials expressed concern about the film's criticism of the industry and its focus on the negative effects on Indigenous people. Though DIAND officials acknowledged that there were going to be greater risks from uranium mines for northern people, they followed direction from the EMR and AECB in their criticisms of the film.

As a supposed protector of Indigenous people, the DIAND was particularly concerned that it was depicted badly. The department wanted to be seen as the advocate of the Wollaston Lake Band in northern Saskatchewan, which led the call for a federal public inquiry on uranium mining after the spill (the DIAND called it a discharge) from the Rabbit Lake mine in late 1989. It's hard to imagine, taking its lead from the EMR, whose minister, Jake Epp, turned down such an inquiry, how the DIAND could act as a legitimate advocate for any Indigenous community.

The DIAND proved to be highly biased towards the pro-nuclear views of its federal counterparts. It continued to misinform northern Dene and Inuit that 50% of the employees were Indigenous at the Cluff Lake mine. While this mine, with the smallest workforce of the three in operation in Saskatchewan in 1990, did achieve 50% northern employment at an earlier point, this category was not exclusively Indigenous. Department officials also wanted to be seen as facilitators of Inuit interests at Baker Lake, N.W.T., when the uranium mine was being proposed. However, their role was to encourage uranium mining as an economic development strategy for the eastern Arctic, and to achieve this they were willing to provide partial or inaccurate information about what has happened in northern Saskatchewan. According to the February 26, 1990, issue of the Yellowknife paper *News/North*, in confidential memos the director general of DIAND's N.W.T. region was predicting the go-ahead of the uranium mine near Baker Lake before the FEARO hearings on this mine had even begun. The DIAND didn't drop its bias toward uranium mining even after the people of Baker Lake voted 90% against the proposed mine. It continued to mimic the views of the industry that radioactivity at the mines is "low," perhaps unaware that a scientific and regulatory dispute continued to rage over the matter. So, while the release of the NFB's *Uranium* didn't expose anyone in the north to radioactivity, the reaction it got fully exposed the pro-nuclear bias of the DIAND and its federal colleagues to northerners.

Notes

1. Canada's occupational limit remains 50mSV or 5 rem. These terms are less confusing when you make the distinction between measures of exposure and measures of dose. Roentgen is about the actual energy from radiation (e.g., 1 roentgen of gamma radiation is about 1 rad, or radiation absorbed dose). Rem stands for roentgen equivalent man, which means a unit of equivalent absorbed dose involving human exposure. (For gamma and beta radiation 1 rad exposure is considered equivalent to 1 rem of dose.) Sivert (mSv) is now being used as an international unit. 1 mSv equals 100 rems. While this is all challenging, the controversies further intensify when trying to estimate the dose-response relationships, i.e., the cancerous consequences of particular exposures and doses. Alpha radiation is particularly controversial.
2. Beak Consultants reported on their contract research on uranium waste management at a CNA conference, International Symposium on Uranium and Electricity, held in Saskatoon, Saskatchewan, in Sept. 1988.
3. Jim Harding, *Aboriginal Rights and Government Wrongs*, pp. 26–30.

chapter 9

DRAWING THE LINE

Private Candu Reactor Proposed for the North

In December 1989, Western Project Development Associates (WPDA), a private consortium, appeared, seemingly out of nowhere, and wanted to build a prototype scaled-down Candu-3 reactor in Saskatchewan's north. With full backing of AECL, the plan was to build a 450 megawatt (MW) reactor with private funds and to sell the electricity to the public utility, SaskPower. Heading the consortium was Colin Hindle, past chair of Saskatchewan's Crown Investments, which oversees the province's public utilities. Another member of the WPDA had been Premier Grant Devine's personal advisor on economic development, and the third was a Tory member on the Native Economic Development Board.

Perhaps the Devine government wanted to float the idea, at arms length, as a means to start privatizing the electricity market. At the time the Mulroney government was looking for ways to expand AECL's meagre sales to make it more attractive for privatization. The proponents used every ploy in the book, even trying to sell the Candu reactor as a good environmental option; as with uranium mining expansion there were promises of an abundance of jobs for a desperate north. However within weeks a coalition of thirty Saskatchewan groups had formed to oppose the plan. The International Uranium Congress, held in Saskatoon the preceding summer, helped solidify relations among non-nuclear groups throughout the province. Along with thirteen major groups, the Congress issued a fourteen-page document "Why We Don't Need or Want a Nuclear Reactor in Saskatchewan," in August 1989.

Many Indigenous groups, including the Métis Society and northern Trappers, expressed stringent opposition to the Candu plan. While NDP Opposition Leader Roy Romanow waffled, Environment Critic Lorne Calvert and the small NDP caucus quickly came out fully in opposition. (Calvert was later to become a pro-nuclear premier of Saskatchewan.) The Saskatchewan Chamber of Commerce was predictably onside, and some town councilors at the town of Big River said they wanted the reactor. Most northern towns did not. Green Lake, which is on the road to the Cluff Lake mine, strongly opposed the plan, and under the leadership of the late mayor and Métis activist Rod Bishop, even sponsored north-south, non-nuclear workshops, called "Fish and Loaves."

Little did non-nuclear activists know that this Candu proposal was part of a concerted effort to create a base in Saskatchewan for the full nuclear fuel system. We were about to enter an overwhelming period of pro-nuclear bombardment, having to resist nuclear advances and propaganda on several fronts.

Tackling Global Warming: Nuclear Not Magic Bullet

The WPDA adopted the CNA assertion that nuclear power was the panacea to global warming. What lay behind this apparent conversion of the nuclear industry to being concerned about the environment? It is noteworthy that no credible international, national or local environmental organization agrees with their claim — simply because there is no scientific basis for it. When the WPDA made its proposal, Komanoff Energy consultants of New York had just analyzed the relative advantages of relying on either energy efficiency or nuclear power to lower global carbon dioxide (CO_2) emissions by 50% by the year 2020 while maintaining a 3% increase in GNP per annum.[1] Komanoff and others found that the nuclear strategy would require reactor plants being built forty-eight times faster than they had been built from 1975 to 1985. Sixteen reactors would have to be built every week from 1995 to 2020. Such a scenario would be completely unecological, due to the massive proliferation of radioactive wastes from the uranium mines, refining and reactors, and it would also be completely uneconomic due to the massive capital that would have to be redirected from society at large. France, the country most committed to nuclear power, was only able to build six reactors in a year during its nuclear heyday in the 1970s, and it has since been forced to cut back to one a year.

In its 1989 report, *High-Level Radioactive Waste in Canada: The Eleventh Hour*, an all-party Parliamentary standing committee called for a moratorium on nuclear power due to nuclear waste buildup. The nuclear waste issue seemed to be the reason why, according to Gallup polling at the time, 70% of Canadians opposed further expansion of nuclear power, and support was down to 16% from 46% a decade earlier. Only three provinces had any nuclear power and Canada only received 13% of its electrical energy from nuclear energy. Rather than further contributing to the growing worldwide problem of radioactive wastes, it was an ideal time to make the transition to sustainable energy policies and to begin to phase-out nuclear energy.

Meanwhile, a revolution was occurring in energy efficiency. For energy efficiency to accomplish the halving of global CO_2 emissions by the year 2020, Komanoff estimated it would only have to increase to an annual growth rate of 4.6%. With few conservation or efficiency strategies yet in place, many countries already had an annual growth rate of 3% or more in energy efficiency in the wake of the oil-price increases and "energy crisis" of the 1970s.

The European Union increased energy efficiency by 20% from the 1970s and targeted another 20% increase by the end of the century. Studies suggested that even the countries most dependent on nuclear power, like France and Belgium, could phase out the industry in four to seven years. In its January 1990 World Status Report on nuclear power *The Economist* reported that with existing energy-saving technology, Britain could cut from 6.5 to 12.5 gigawatts (GW) of the country's 55 GW capacity by the year 2000.

The U.S. Rocky Mountain Institute calculated that money spent on energy efficiency is up to seven times more effective at reducing CO_2 than money spent on nuclear plants. According to Lawrence Berkeley Labs, installing cost-effective lighting throughout the U.S. would eliminate the demand for electricity equal to that produced by the forty 1,000 megawatt (MW) nuclear reactors operating in the U.S. The American Council for an Efficient Economy estimated another 40,000 MW could be saved by introducing the most efficient refrigerators, air conditioners and water heaters.[2]

The nuclear industry wants the public to believe that coal plants are the major cause of the greenhouse effect and that nuclear plants are the alternative, but a Swedish study discussed in the December 1988 *Energy Policy*, poked holes in this attempt to get on the environmental bandwagon. It pointed out that if all nuclear reactors were shut down and replaced by coal plants, this would add only 0.1°C, or 1/200th of the estimated 2°C increase in global temperature predicted by some climatologists by the year 2025.

Komanoff's research shows "that efficiency is well poised, while nuclear power is not, to ameliorate the Greenhouse effect." But there isn't going to be a simplistic solution to the global warming crisis. Besides implementing efficient and renewable energy systems, a successful sustainable strategy will need to stop other atmospheric pollutants, including those damaging the ozone layer, stop the destruction of the rain forests, which are critical to the carbon cycle, and quickly develop an alternative to fossil-fuel-burning transportation.

But the backers of a Candu plant for the north didn't want to hear any of this. They were misinformed, confused or dishonest about the greenhouse issue. They persisted with the view that nuclear power is environmentally safer than other forms of energy, failing to mention that there had been a massive nuclear accident at Chernobyl just three years previously. This accident released many times the atmospheric radiation coming from the Hiroshima bomb and led to twenty-seven cities and villages being abandoned and the mass slaughter of 90,000 reindeer in Sápmi. Estimates of additional human cancers caused by this reactor accident at the time ranged as high as one-quarter of a million people, and this may prove to be an underestimate.

The Candu proponents made no mention that many reactors have already been shut down permanently due to serious safety problems. They never mentioned that nuclear power is highly inefficient — for example, it generates massive waste heat at the plant before transmitting electricity that then gets used to create heat miles away. They made no mention that the nuclear industry is unable to get liability insurance due to the catastrophic environmental and public health costs of a major accident. For this reason their proposed private plant would still require that the government, and thus the public, take the economic and legal risks. In Britain, where Prime Minister Thatcher tried and failed to privatize the state nuclear electricity industry, the public was still to pay much of the costs of decommissioning, fuel-reprocessing and fuel "disposal" as part of the conditions of sale.

If predictions by the nuclear industry about reactor safety had been true, there would only be a major accident every 2,000 years (assuming 500 reactors in the world). And these ridiculous estimates seriously downplayed consequences by ignoring delayed deaths, i.e., latent cancers, in their calculations. In fact, there have already been major accidents at Windscale (1957), Three Mile Island (1979) and Chernobyl (1986), and some now predict major accidents every decade or two. Nor is the Canadian Candu reactor safety record squeaky clean. There were accidents at the Chalk River nuclear reactor in 1952 and 1958 and at the Pickering plant in 1983. The Bruce plant had equipment failures and errors in 1979, which, if they had occurred simultaneously, might have led to a catastrophe on the scale of Three Mile Island.[3]

There was confusion about the role of government in this Candu proposal. While Premier Grant Devine expressed qualified support for a private nuclear reactor, he also said that utilities like SaskPower and SaskTel should remain public corporations. However, suspicion that he wanted to privatize Saskatchewan's utilities, as well as widespread scandals, would eventually bring his government down. Though the Candu plant was promoted by a private group, a WPDA representative said SaskPower would choose the site, and he called for the same federal financing that would be received by "public" utilities.

In giving his support for a private nuclear reactor, Premier Devine called uranium "nature's" or "God's gift"[4] — a fundamentalist view that the world is here for our dominion and for our taking, often held in contrast to the notion of stewardship. However, while elemental uranium exists in a natural ore (where many of us believe it should stay), plutonium — the extremely toxic transmuranic element created by nuclear fission with a half-life of 24,400 years and used in nuclear weapons — is not a gift of nature or God, but an unwanted "gift" from the nuclear industry. How quickly Premier Devine forgot that he campaigned and won the 1982 election using pamphlets saying "The price of uranium has fallen by more than one-half. Investment in uranium doesn't help today and it does not secure our future tomorrow."

Premier Devine's approach to the environment proved to be inconsistent and somewhat contradictory. His government had chosen not to put state-of-the-art flue-gas scrubbers in the Shand coal-fired electrical plant, which could have removed 90% of the sulphur dioxide (SO_2) being emitted from its stacks. As a consequence, the plant barely met existing federal emission standards and was destined to undermine the Kyoto Accord. The shallowness of the Premier's environmental awareness came to light when he downplayed the problems of acid rain in rural Saskatchewan by saying it might actually be good for our low-acid soil. This simple-mindedness made it difficult for the Premier to openly jump on the bandwagon that was erroneously promoting nuclear energy as a panacea for fossil fuel problems such as acid rain. It was, after all, Devine's government that shut down SaskPower's Office of Conservation in 1982 and appointed management singularly committed to the market view of public utilities, a view that stresses increased energy consumption and sales as a means to reduce debts incurred by high capital costs — which in turn are required due to growing demand and lack of conservation. This circular and unsustainable approach benefits only those who live off capital.

Rejecting "Jobs at Any Cost"

The backers of a private Candu tried to entice people with the prospects of new job opportunities, and this "jobs at any cost" approach had some appeal with the economic downturn and prairie drought of the 1980s. However, over $600,000 capital investment for each twenty-year job (equivalent), without even considering inflation and cost overruns typical in such mega projects, is hardly the way to tackle unemployment rates of up to 80% in some northern areas.

According to the U.S. Federal Energy Administration the ratio of trades personnel to scientists in a solar energy economy would be 9:1 compared to only 2:1 in a nuclear economy.[5] Many northerners, most notably the Métis, who had witnessed the failure of the uranium industry to deliver the jobs promised to northern Indigenous peoples during the public inquiries a decade previously, un-categorically opposed the proposed nuclear plant.

Concern for the environment, demand for energy and employment in the north proved to be trumped-up rationalizations for this proposed nuclear plant. With the non-nuclear movement militantly contesting the nuclear propaganda, the proposal failed to get sufficient support in the court of public opinion to even get to the review process. This made things even worse for the AECL. By 1989 the Canadian nuclear industry had sucked up an estimated $12 billion in public subsidies just to stay afloat. It was aware that it might not be able to continue to expect such preferential treatment because of the federal government's concern with the fiscal crisis. And it had failed over the previous decade to export the Candu to Third World nations. By the time of this proposal, the AECL hadn't made any sales since its 1981 sale to Romania, with none occurring until its sale to South Korea in late 1990. The international market for nuclear power was rapidly shrinking, and AECL was starting to compete with the much larger Framatome of France, Westinghouse of the U.S. and Kraftwerk Union of West Germany. Its attempts to manipulate growing public concern about chemicals in our food chain into support for irradiated food[6] or to diversify into nuclear waste "disposal" were not enough for it to survive. Thus it hoped a scaled-down 450 MW Candu 3, built in the north, would act as a showpiece for a new export marketing strategy to attract industrializing Third World countries.

This trial balloon was the first of many attempts to get a reactor built in Saskatchewan. It burst almost as soon as it was floated, with the vast majority of Saskatchewan people indicating they wanted no part in nuclear power. While many people remained uncertain and confused about uranium mining, perhaps wanting to believe it was just another resource industry creating jobs and bringing royalties to the provincial coffers to pay for health care, the line had been drawn over nuclear energy. Stopping this proposal so quickly was the second big victory, after stopping the uranium refinery at Warman, of the Saskatchewan non-nuclear movement.

Universities Willing Partners in Candu Strategy

Though the people of Saskatchewan had no need or desire for nuclear power, the professional and business elites who would directly benefit persisted with their lobbying. Saskatchewan's two universities became launching pads for the proposed Candu reactor industry in the 1990s, much as they were willing partners in the expansion of uranium mining in the late 1970s. I was on the University of Regina faculty during this period and observed the pro-nuclear bandwagon first-hand. In the 1980s I undertook extensive "Human Context of Science and Technology" research on the uranium inquiries, for which I got funding from the Social Science and Humanities Research Council.[7] However, when a seminar was called for faculty doing research on uranium, I wasn't invited. When I heard about the event, I insisted on being included on the agenda, citing academic freedom. Other presenters doing contract research in geology, radioactivity measurement and other nuclear-related areas took it for granted that university research went hand-in-hand with the uranium industry, while I presented on the contentious historical, economic, environmental and legal issues that were involved in Saskatchewan uranium policy.

Professor Bev Robertson, who organized this seminar, later wrote opinion pieces published on November 6 and December 4, 1989, in the Regina *Leader Post* advancing the view that the nuclear conflict was between expert scientists and illiterate activists. Another physics professor caught up in nuclear politics was Len Greenberg, who took some very silly public positions, reported in the December 30, 1988, *Leader Post*, such as the only environmental effect from the warmer water created from the cooling of a Candu plant would be bigger fish. A common-sense criticism of Greenberg was published under the headline, "Nuclear logic defies belief," in the January 16, 1989, *Leader Post*. I once attended a lecture by Professor Greenberg, where he was clearly unaware of the research in health physics on low-level radiation hazards or the scale of the Chernobyl accident.

Later, some academics in the Faculty of Engineering and Department of Computer Sciences, wanting to reap the benefits — including from lucrative reactor design research — began to show interest in the proposed Candu development. Pro-nuclear guests regularly attended engineering classes, and some students came to me privately to seek access to other more balanced information. Then in March 1989 the Faculty of Science embarrassed itself by bringing in a crass nuclear ideologue to give its prestigious Basterfield Lecture.[8] There seemed no limit to the extent to which the cash-strapped university and some grant-craving academics would prostitute themselves. When the Faculty of Engineering invited the head of the AECL to speak at an academic lecture series in March 1990, it was time to bring an alternative viewpoint to the seemingly lethargic "community of scholars."

The AECL's president spoke at the Engineering lecture series entitled "Issues in Technology," which was "to focus on the major technological issues which are transforming our society." The event, however, was more promotional than educational and was co-sponsored by business groups, such as the Regina Chamber of Commerce and some engineering firms that hoped to benefit from a Candu industry. This nuclear power event was held at the Saskatchewan Trade and Convention Centre, with a $40 entrance fee to hear the AECL essentially promote itself. An educational event befitting the university's non-partisan, academic role would have been an open event, on campus — not a corporate event with a restricted audience. If education had been the primary motivation, a scholar examining from a detached perspective how the nuclear industry was transforming society would have been invited to take part. Or, if having an AECL spokesperson was paramount, why not also sponsor a speaker who could provide an alternative view? This would have shown more respect for the intelligence of the audience and the right of citizens to be presented adequate and balanced information with which to make up their own minds.

Some of the issues that weren't discussed by the AECL lecturer include higher childhood leukemia rates around nuclear reactors;[9] higher childhood diseases among children of nuclear plant workers; the legacy of spills and buildup of long-living radioactive tailings at Saskatchewan's uranium mines;

and the cost-effectiveness of energy efficiency compared to nuclear power for addressing global warming. And there definitely was no mention of the steadily rising fatality rates after Chernobyl.

The CNA's 1987–88 Business Plan revealed a strategy to establish better relations with supportive people in universities, schools and the media. Millions of dollars had already been pumped into government-subsidized high-pressure ads and the insertion of pro-industry resources into our public schools and universities, while under-funding of public education was blamed on the federal deficit and rising debt. There were already several indications of links between elites in the uranium and nuclear industries and Saskatchewan's two universities. John Nightingale, who was chair of the Board of Governors at the University of Saskatchewan from 1989–92 was also a past president of both the Key Lake Mining Corporation and the Saskatchewan Mining Association. At the time, the Board was considering an AECL offer to build a new Slowpoke reactor on campus.[10] Sylvia Fedorak, a past chancellor of this University and a long-time proponent of uranium mining, had also been a member of the AECB. She continued with her pro-nuclear promotions even after she was appointed lieutenant governor of Saskatchewan. Long-time Regina lawyer Derril McLeod was chair of the University of Regina Board of Governors from 1974–80, during which period he was also legal counsel for the French uranium company Amok during the CLBI and did some Amok negotiations with the SMDC. Later, from 1983–89, he became chancellor of this University. The nuclear industry worked strenuously to establish such inside tracks, but why did our universities with all their interdisciplinary brainpower so easily accept this shallow and opportunistic ploy?

The title of the AECL president's talk was "Nuclear Power: The Clean Air Alternative to Sustainable Energy," which was likely intended to reinforce the nuclear industry's costly advertising campaign promoting itself as the panacea to greenhouse gases and acid rain. Calling radioactive gases "clean air" distorts our language and knowledge, as radiation being invisible to our senses doesn't make it clean or harmless. The skepticism rightly expected in a university climate regarding such manipulation of language was noticeably missing. Why, for example, when AECL was calling nuclear "clean" and downplaying the Chernobyl accident of April 26, 1986, to promote its Candu strategy, were some of my science-background university colleagues so silent on this matter?

In her 2006 book, *Nuclear Power Is Not the Answer*, Helen Caldicott shows us why the Chernobyl accident can't be downplayed. Due to an unethical agreement in 1959 between the United Nations agencies, the International Atomic Energy Agency (IAEA) and World Health Organization (WHO), the WHO was muzzled from assessing the dire health consequences of this or any nuclear catastrophe. Luckily other U.N. agencies have spoken out. In

1994 the U.N. Office for Coordination of Human Affairs estimated that 8.4 million people were exposed to radiation, that 400,000 people had to be relocated, that 150,000 square kilometres were contaminated and 52,000 square kilometres were ruined. In 2001 the United Nations Development Program released a report that talked of the epidemic of trauma as well as cancer. Caldicott says that as many as 10,000 people in the clean-up crew have died prematurely, 20% of Belarus and 8% of Ukraine was contaminated and radioactive "hot spots" exist as far away as Britain and Sweden. She discusses how the highly nuclear-dependent country, France, tried to deny the contamination on its own land and people. Caldicott reports that from 1986–2001 there were already 8,000 excess childhood thyroid cancers in Belarus alone. Long cancer latency periods mean that much more human suffering is to come.

The State Centre of Environmental Radiochemistry in Kiev calculates the radiation levels in the principle river in Ukraine, the Dnieper, will not peak for sixty to ninety years. Peak cancer levels are not likely for three decades, and some independent scientists project as many as one million long-term fatalities from Chernobyl's radiation.[11]

In spite of the Chernobyl catastrophe, nuclear misrepresentation continued full steam ahead, with the industry selling itself as the alternative to pollution from coal-fired electrical plants. This is highly opportunistic, as carbon from coal contributes only about 10% of the total greenhouse gases. Carbon from all fossil fuels used to generate electricity contributes about one-sixth of these gases. The cost of changing from fossil to nuclear fuels for electrical generation, an estimated $8 trillion, according to the work of Charles Komanoff, would be astronomical. The scenario is ridiculous, uneconomical and unnecessary, since, as we saw earlier, the same effect can be achieved much more cheaply by increasing energy efficiency by 5% per year.

To avert a global warming disaster, massive changes are urgently required in forestry, transportation and agricultural practices, in addition to energy production, since these contribute the bulk of greenhouse gases. Methane coming from the livestock industry contributes substantial greenhouse gases, and meat-producing agribusiness in turn perpetuates deforestation for grazing, which engenders an ecologically unsustainable vicious circle. Saskatchewan universities are well placed to undertake interdisciplinary research on these matters. Why then did they cozy up so closely with the nuclear industry and its intellectually and scientifically flawed agenda?

Nuclear Energy Not Sustainable Development

There are no simple tech-fixes to save the planet from industrial pollutants and global warming. A change in overall direction is urgently required.

However, the global challenge of sustainability was crassly misrepresented by the university's nuclear proponents. The glossy advertising for the AECL guest speaker incorporated "green" symbolism to appeal to the public's growing environmental concerns. A closer look at this symbolism reveals an underlying technocratic — not ecological — viewpoint. The "arm" of a robot looks like it is picking a green apple, which seems to suggest reconciliation between nuclear technology and environmental protection, a kind of nuclear totem. The sub-text is that nukes are environmentally friendly, like green products promoted in some supermarkets.

But why would a robot be picking an apple in the first place? Robots are often used in the nuclear industry because of the threat to humans from direct exposure to nuclear fuel or waste. Robots may have to be used to mine the high-grade uranium at the proposed Cigar Lake underground uranium mine. In a world of expanding nuclear energy such as the AECL is promoting, radioactivity would inevitably continue to enter into the earth's ecosystems from the uranium tailings left at the mines and the nuclear releases and wastes from the reactors. The graphic can therefore be interpreted as demonstrating that, in such a nuclearized world, an apple (symbolizing the food chain) might become so contaminated that humans would not be able to safely pick or eat it.

Sometimes self-interest can get caught up in its own rhetoric. The title of the talk revealed how the AECL and its university sponsors thoughtlessly tossed in the much-used term "sustainable" for effect. The sub-title "Alternatives to Sustainable Energy" is inherently confusing. Whether an unconscious slip and/or the kind of oversight that occurs when people use PR clichés without any underlying substantive knowledge, the inadvertently revealing title of the AECL talk implied that nuclear energy is not sustainable. This was clearly not the intention. However, with its toxic radioactive reactor emissions and wastes and the ecological damage left at Saskatchewan's (and Ontario's) uranium mines, nuclear energy is, in fact, not ecologically sustainable. And this is true even without catastrophic accidents like Chernobyl.

Incredibly this promotion came out of an institute of higher learning. The use of the term "sustainable" was likely an attempt by the AECL to hitch its nuclear wagon to the groundswell for sustainable development, after the United Nations' World Commission on Environment and Development coined the term in 1987. But there are some serious problems with the way the term is being mainstreamed. David Suzuki put it succinctly back in 1989, in his column in Regina's *Prairie Dog* magazine, when he said

> The phrase sustainable development seems to offer the hope of both preserving the environment and continued growth and development. It is a terrible delusion that distracts from the inescapable

> consequence of exponential growth — once the explosive phase of increase starts, it cannot continue and will stop. The only question is how — by war, famine, disease or deliberate planning.

The AECL and its university supporters didn't seem to know that the U.N. report on sustainable development did not promote the nuclear industry. The report, known as the Brundtland Report after its chair Gro Brundtland, was published by Oxford University Press in 1987 under the title *Our Common Future.* In its discussion of nuclear power on pages 182–89, it acknowledged that nuclear energy produces only 15% of global electricity (which is only about 3–5% of primary energy) and "has not met earlier expectations that it would be the key to ensuring an unlimited supply of low cost energy." Furthermore, it concluded that "costs... have increased more rapidly for nuclear stations during the past 5–10 years, so that the earlier clear cost advantage of nuclear over the service life of the plant has been reduced or lost altogether." It also noted that "nuclear provides about one-third of the energy that was forecast for it 10 years ago." Although the Brundtland Report acknowledged the attempts to control the proliferation of nuclear weapons by trying to separate nuclear weapons and nuclear power, it admitted that "for countries with full access to the complete nuclear fuel cycle, no technical separation really exists" and "thus there still remains a danger of proliferation of nuclear weapons." The Commission was quite emphatic about this, saying, "The Non Proliferation Treaty has not proved to be a sufficient instrument to prevent the proliferation of nuclear weapons."

The Brundtland Report also noted that the nuclear industry (this includes the Canadian industry) has been claiming that the chance of a major reactor accident was one in a million years of reactor operation. But the Three Mile Island accident in 1979 and the Chernobyl accident in 1986 occurred much more often, "after about 2,000 and 4,000 reactor-years respectively." The report continued, saying the risk of such accidents "is by no means negligible for reactor operations at the present time." Finally, the Brundtland Report refers to the "many thousands of tons of spent fuel and high level waste" and the need to isolate these "from the biosphere for many hundred of thousands of years that they will remain hazardously radioactive." On this matter, which the nuclear industry tries to convince the public is a non-issue which should be left to their "experts," the Commission was unambiguous, saying "the problem of nuclear waste disposal remains unsolved."

One wonders whether the AECL or its sponsors within the hallowed halls of academia ever looked at the Brundtland Report, which concluded that "The generation of nuclear power is only justified if there are solid solutions to the presently unsolved problems to which it gives rise." In view of these serious unsolved risks it is not surprising that many countries were already adopting non-nuclear policies and strategies for phasing-out nuclear power. The Brundtland Report set out three broad policy options regarding nuclear energy. Several countries decided to take the first option and "remain non-nuclear and develop other sources of energy." At the time, most of Canada had already gone this route, and we must ensure that Saskatchewan joins this "non-nuclear" club. While Ontario cannot go this route immediately, its most ecologically responsible path would be to follow the second option and see its "present nuclear power capacity as necessary during a finite period of transition to safer alternative energy sources."

Meanwhile, propagandizing us with our own tax dollars, the nuclear industry understates its costs and risks, falsely calls itself "clean" and tries to convince citizens to take the third option and "adopt and develop nuclear energy with the conviction that the associated problems and risks can and must be solved." This is a leap of faith Canadians cannot afford to take.

The Brundtland Report recommends

> that vigorous promotion of energy-efficient practices in all energy sectors and large-scale programmes of research, development and demonstration for the safe and environmentally benign use of all promising energy sources, *especially renewables, be given the highest priority.* (emphasis added).

The chair of the World Commission, Gro Brundtland, stressed the same point in a guest essay in the September 1989 issue of *Scientific American*. We can all be thankful that Gro Brundtland has returned to global public service as the U.N.'s envoy on strategies to reduce global warming.

The more that scarce resources are committed to creating new energy supplies, whether nuclear or heavy oil, the less will be available to create energy efficiency, conservation and renewables. A shift towards sustainable energy requires that we quickly bring truly ecologically friendly renewable energy systems on-stream. Were the huge public subsidies that go to the AECL ($140 million the year this university event occurred) to be redirected towards developing wind, solar and other renewables, these would be much more quickly marketable. Instead, research and development grants for alternative, non-nuclear energy technologies were further cut by the federal government of the time.

Corporate, not ecological, interests — including in our universities — unfortunately dominate research and development decisions. At the time of this AECL offensive in Saskatchewan, about $20 billion went worldwide, annually, into new electrical generating capacity, while only $100 million went into conservation. The worldwide costs of constructing nuclear power plants to replace fossil-fuel plants were estimated to be about $8 trillion, and these would only address a small amount of the total sources of greenhouse gases. Seven times the reduction of these gases could be obtained per dollar through applying scarce capital towards energy efficiency.

This AECL-backed proposal for a Candu in Saskatchewan was a wrong-headed and wasteful non-starter that relied on a massive public relations campaign of disinformation. The dynamics of the controversy tell us much about the politics of nuclear expansion that will help us in the years of struggle that lie ahead. The AECL's technical and business allies wanted to share in the spoils; that some segments of the university joined the ranks of uncritical cheerleaders seriously eroded its image as a place of higher far-thinking learning.

Notes

1. Charles Komanoff, "Greenhouse Effect Amelioration-Efficiency vs. Nuclear," Komanoff Energy Associates, New York, memo, Aug. 24, 1988. Also see Bill Keepin and Gregory Kats, "Greenhouse Warming: Comparative analysis of nuclear and efficiency abatement strategies," *Energy Policy*, Dec. 1988, pp. 538–61.
2. See Rocky Mountain Institute and other findings in Bill Keepin and Gregory Kats, "Greenhouse Warming," *Energy Policy*, Dec. 1988, pp. 538–61. The potential of energy conservation was also discussed by Bill Harding, "Energy efficiency needs priority," Regina *Leader Post*, Jan. 25, 1989, p. A7.
3. More details on realistic calculations of reactor accidents are given in Jim Harding, "Madness to increase nuclear mishap odds," Saskatoon *Star Phoenix*, March 3, 1989.
4. "Uranium called a gift," Regina *Leader Post*, Jan. 10, 1989, p. A4.
5. Environmentalist for Full Employment and Fast Solar Energy Facts regularly update information on jobs from solar versus nuclear energy.
6. For a critique of food irradiation, see Tony Webb, Tim Lang and Kathleen Tucker, *Food Irradiation: Who Wants It?* Rochester, Vermont: Thorsons Publishers, 1987.
7. Several reports were published by Prairie Justice Research, under the Series "In The Public Interest."
8. See "Nuclear power causes needless worry: Professor," Regina *Leader Post*, Mar. 8, 1989, p. A3; and "Basterfield bastardized," Univ. of Regina, *Carillon*, Mar. 16, 1989, p. 15.
9. See "Leukemia study to start soon," Regina *Leader Post*, Oct. 2, 1990, p. A8; and "Childhood Leukemia increases around nuclear facilities," *The Anti-Nuclear Review*, Summer/Fall, 1989, p. 3–4.
10. See "Investigation of The Slowpoke Energy System," memo, Acting President, University of Saskatchewan, October 17, 1989; also "Province may fund reactor," Regina *Leader Post*, Jan. 24, 1990.
11. "UN Pleas for Chernobyl Victims," April 26, 2001, available at <http://news.bbc.co.uk/2/hi/europe/1297261.stm> accessed June 2007. Also see linked stories.

chapter 10

COMMON SENSE ABOUT NUCLEAR WASTE

Accumulating nuclear waste (spent fuel) remains a major Achilles' Heel of the nuclear industry, and the AECL has always known that tackling this problem was indispensable to any strategy of expansion. In 1990 the Federal Environmental Assessment Review Office (FEARO) set up an Environmental Review Panel to hold hearings across Canada on the AECL's proposed deep rock nuclear waste storage. Of all the locations where hearings were scheduled, Saskatchewan was the only one without nuclear reactors or an active nuclear waste storage project, which raised the question of motive: was it because of the AECL's proposal to build a Candu-3 in the province, or because the AECL was planning to target the north as a potential nuclear spent fuel site — or both?

By then, those of us working for a non-nuclear future had little confidence in the public inquiry process. Nevertheless, when the Environmental Review Panel came to Regina on November 19, 1990, I reluctantly made a presentation on behalf of the International Uranium Congress (IUC). Somewhat to my surprise, some members of the Panel were interested in our presentation and wanted to continue in conversation after the scheduled session was over. Lois Wilson, a Panel member and past moderator of the United Church, went on to publish a United Church book on the issue, *Nuclear Waste: Exploring the Ethical Dilemma*.

At the time the Regina-based IUC animated an international network of environmental, peace, development and energy organizations in uranium producing and consuming countries. Our broad non-nuclear objectives were endorsed by over 160 organizations from twenty-two countries, including eight Canadian provinces and territories and sixty organizations from Saskatchewan.

I began by asking why Saskatchewan was included in this review. I had been told earlier, at an open house sponsored by the FEARO, that we were included because we have uranium mines, and later in a phone inquiry, that it was because there had been a proposal to build a nuclear reactor here. I pointed out to the Panel members that their own terms of reference said they weren't to be site specific, but this didn't mean that their being in Saskatchewan couldn't be a positive development. Some of us had been calling, for more than a decade, for a federal public inquiry on the whole nuclear fuel system, and perhaps this review might be a step in that direction. Their coming here to consider what to do about the steady buildup of nuclear reactor wastes might even convince the few groups still endorsing a nuclear reactor for Saskatchewan to reconsider their position.

I presented the view that no technology, nuclear or otherwise, without a proven waste management system should be developed in the first place. This is one of the many implications of thinking through the matter of environment, development and sustainability, sparked by the Brundtland Report in 1987. In the case of the nuclear power industry, which has been in existence for more than a half century and continually failed to fulfill its guarantee of a waste storage system, the only socially and ecologically responsible thing to do is to immediately stop using the technology, thus preventing a buildup of even more nuclear wastes — a cost and curse for future generations.

The Federal Standing Committee on Environment and Forestry had already made a call for a moratorium on nuclear power in Canada in 1988, and such a call was also made in 1978 by Ontario's Porter Commission on Electric Power Planning in its report, *A Race Against Time*. I argued that it was long past the time to heed this call. Even without considering further catastrophic reactor accidents, the continued proliferation of nuclear weapons and the accumulation of uranium tailings in the north of Saskatchewan, the risk from the buildup of nuclear wastes alone would justify the phasing-out of nuclear power altogether.

If the Panel was brought to Saskatchewan in part because we have the world's largest uranium mines, then why, I asked, wasn't the handling or mishandling of radioactive wastes coming from these uranium mines included in the review. There were already nearly 200 million tonnes of radioactive tailings built up at Canadian uranium mines — tailings that contain the vast bulk of the radiation from the natural uranium-bearing ore. Furthermore, because of the particularly high percentage of uranium in the ore being mined in our north (e.g., the ore at Cigar Lake averages about 12–14%), the tailings from uranium mines in Saskatchewan are particularly radioactive. A strong case can therefore be made that the high levels of thorium-230 and radium-226 in these tailings should qualify as high-level wastes. And the only reason these high-level tailings exist is because of nuclear power (and nuclear weapons); hence it was rather academic to pretend that the issues could be split off so neatly from each other.

The last time the FEARO created a panel in Saskatchewan, in 1980 to deliberate on a proposed uranium refinery near Saskatoon, it too refused to include uranium mining in its considerations. Since there could be no refinery without uranium being transported from the mines, this puzzled environmental and public interest groups at the time. But at least that panel was in Saskatchewan because an actual facility, a uranium refinery, was being proposed.

I nevertheless encouraged the Panel to have the AECL pursue, in its EIS, the degree of congruency, if any, of actual practices used with uranium tailings with the ones presented and supported in public inquiries. Specifically, did what was proposed at the Cluff Lake, Key Lake and Rabbit Lake mines bear any relation to what was actually done with the tailings at those sites? For example, the high-level radioactive tailings, which were to be isolated from the other tailings at the Cluff Lake mine, ended up leaking from the cement containers, which were to be buried, and were consequently reprocessed and put in with the lower-level tailings. This option of adding high-level wastes to the tailings was soundly rejected in the original review, yet this is what happened.

This was pertinent to the FEARO Panel's review because it is a significant test-case of how the front-end of the nuclear industry actually performs, in contrast to what it claims will be its "state of the art" fail-safe waste management methods. The Panel therefore had an opportunity to critically assess the value of basing waste management plans on idealized models devised by the industry.

I also encouraged the Panel to ask the AECL to include in its EIS a complete record of all past attempts at nuclear reactor waste "disposal" worldwide, with a full record of the outcomes. Several countries have tried and abandoned deep rock disposal, for a variety of geological and ecological reasons, not the least being the movement of underground waterways. Such movement of underground water has already occurred in Canada's nuclear reactor waste storage experiments done near Lac du Bonnet in Manitoba.[1]

Everywhere there has been nuclear power, authorities remain confounded by what to do with the wastes. The depths to which the nuclear industry will go to pretend to solve its radioactive waste problem is revealed by the U.S. plan, taken seriously until quite recently, to "dispose" of its hundreds of thousands of tonnes of reactor waste by torpedoing them into the seabed under the North Sea. It was claimed that the seabed would encase the radioactive elements and therefore slow down their diffusion as the encasement eroded. Another proposal was to shoot the nuclear wastes into space. These are just a few of a series of astonishing "solutions" — with no ecological merit — that the nuclear industry has seriously considered. And such astonishing "solutions" continue to this day. Bill Adamson notes that, "Germany has passed legislation to down-phase and gradually close down its nuclear reactors. It tried storing high-level waste in underground salt mines at Moresleben. Now huge salt blocks are falling from the shaft's ceiling. To prevent total collapse, authorities are proposing to fill the shaft with a special concrete, 4 million cubic feet, taking 15 years to complete, at a cost of $3 billion CDN."[2] In the U.S., officials now look to more arid locations for possible storage. Their present "plan" is to bury nuclear wastes at Yucca Mountain in Nevada after

2017, but in order to go ahead the Department of Energy must show that there will be no radioactive leakage for 10,000 years. Under growing pressure, the Bush Junior administration attempted "interim" storage, but this has not met with public support. A September 20, 2006, press statement by the Ohngo Guadadeh Devia Indigenous people indicates that a grassroots group, Indigenous Environmental Network, helped defeat a plan for interim storage on the Skull Valley Goshute reservation in Utah, one of the poorest reservations in the country. The federal government has now gone back to Nevada with millions of dollars to convince the state to agree to an interim storage site at Yucca Mountain.[3]

Energy Policy Always Below the Surface

If the FEARO held hearings in Saskatchewan in part because of the AECL proposal for a nuclear reactor, they had the cart before the horse. There was no actual project, no site and no environmental review process in place. And there was widespread opposition to any nuclear reactor, including redesigns of the Candu or Slowpoke. The Official Opposition at the time, the NDP, was on record as opposing the proposed nuclear reactor, and any provincial government publicly endorsing such a reactor would likely guarantee its own defeat. As the public response to the proposed uranium refinery near Warman showed a decade previously, deep opposition to nuclear power exists in rural and urban areas, and Grant Devine's support for a private Candu contributed to his shrinking support and eventual electoral defeat in 1991.

That the FEARO Panel came to Saskatchewan because a nuclear reactor had been proposed seemed a strange rationalization. According to its terms of reference, the Panel was to exclude "energy policies of Canada and its provinces, and the role nuclear power should play in those policies,"[4] and they had already used this mandate to justify excluding such related issues as reactor safety, nuclear fuel reprocessing and the military aspects of the nuclear industry. However, an inquiry like this, especially when it wasn't site specific, couldn't pretend to avoid basic energy policy questions. The question of nuclear wastes cannot be addressed without looking at how, where and why they are created, which is why several respected and responsible bodies had already called for a moratorium on nuclear power until, at the very least, the question of storage could be adequately addressed. This is inherently an energy policy issue.

The enforced separation of waste management from waste production clearly works in the interests of nuclear industry propaganda. A full inquiry into nuclear wastes must consider whether there was an alternative to this waste production in the first place. Had the Panel desired to play a constructive role regarding nuclear waste buildup and the reduction of future hazards, it would have inquired into energy alternatives such as efficiency, conservation and renewables, and how their development could reduce the use of nuclear power and the buildup of nuclear wastes. This preventive approach has become the preferred approach taken to other waste problems, such as PCBs, dioxins, methyl mercury and asbestos, to name a few. In cases such as nuclear power, where the toxic wastes are inherent to the industry, an alternative technology is imperative. This common-sense approach seems to be lacking in the promotion of nuclear energy.

Nuclear industry officials try to market their toxic product as the sustainable alternative to fossil fuels, but even a brisk reading of the United Nations Brundtland Report indicates that nuclear energy is not the way to go. In the previous chapter, I noted that the report concluded, "The generation of nuclear power is only justified if there are solid solutions to the presently unsolved problems to which it gives rise." While this could be interpreted as not an outright rejection of nuclear power, there's no doubt that renewable sources are given the highest priority by the United Nations Commission. On behalf of the IUC, I therefore encouraged the Panel to ask the politicians and bureaucrats in the federal Ministry of Energy, Mines and Resources (EMR), which subsidizes the AECL, why they weren't pursuing energy priorities emphasized by the Brundtland Report. By then the federal government, like many other levels of government, had paid much lip service to "sustainable development," but the burning question was: when was it going to act on it?

There is growing worldwide support for the ecological ethic that wastes for which a safe disposal system is not in place and is not likely to be developed should simply not be created. Since nuclear wastes will remain hazardous to the biosphere for hundreds of thousands of years, one cannot really talk of disposal as a viable waste management strategy. One is really talking about storage in perpetuity in an environment that is ecologically and geologically dynamic and unpredictable, for time periods beyond human comprehension. To put this in perspective, the management of plutonium for the required 800 generations is five times the time span that it has taken humans to expand out of North Africa and "colonize" the whole planet.

At the time of the Panel, it was estimated that there were over 40,000 tonnes of high-level nuclear wastes stockpiled in Canada alone. Nuclear proponents commonly play the quantification game and argue that the volumes of nuclear waste averaged out across those using electricity is quite low. Actually they should average it for those using nuclear-generated electricity and compare it to wastes from renewable sources of generating electricity, like wind and solar energy. The main issue, however, is qualitative not quantitative. When a nuclear proponent cleverly tells me the nuclear waste volume averaged out for consumers would amount to only an aspirin, whether or not the calculation is technically correct, I respond with: would you rather take an aspirin size tablet of acetylsalicylic acid or of plutonium?

It's hard to fathom the qualitative dimension of the toxicity of nuclear reactor wastes. Neither the poisonous fission products (iodine-131, strontium-90, or cesium-137) nor the poisonous transuranic elements (plutonium, neptunium, americium and curium) would exist without nuclear power plants. The new element plutonium, never known to planet earth before nuclear reactors, is one of the most toxic threats to future eco-systems and inhabitants. With a half-life of 24,400 years, even after a quarter of a million years of radioactive decay, a tonne of plutonium would still remain as a kilogram of plutonium, which, if dispersed, would still be enough to poison a billion people.

If Canada's nuclear wastes were diluted to meet the maximum levels of radiation legally allowed at present, they would probably contaminate all the earth's oceans and lakes twice over. While such a scenario of dilution isn't likely, the long-term ecological threat should be taken very seriously. With all the current sources of nuclear wastes and a continuing buildup, there is a high likelihood of its ecological dispersion posing a threat to future generations.

With the stakes so high, the burning question is how to prevent the production of nuclear wastes. The answer is obvious and quite simple and should not be avoided whatever the political or economic diversions. In support of such common sense and looking at the source of nuclear wastes, no government inquiry to date has yet considered separating a review of plans for dealing with uranium mine tailings from a review of the uranium mines that produce them. This was true for the CLBI, and the FEARO review of the Kiggavik uranium mine proposed and rejected near Baker Lake, N.W.T., included the issue of uranium tailings management. The artificial and questionable separation of the review of nuclear wastes from the nuclear energy policy that creates them, treats nuclear power as a sacred cow with privileged status. The federal government seems so tied to its offspring, nuclear power, that it cannot take a second look — even in its own environmental review process. The desire to maintain the technology for joint military purposes with the United States cannot be ruled out as a main reason why.

Nuclear Bias in Federal Environmental Review Process

It is pretty clear why there is widespread public concern about the pro-nuclear bias so ingrained in the federal government, which heavily subsidizes this technology, compromising the independence of the review process. Such a bias was clearly shown in the narrowness of the Panel's Scientific Review Group: six of thirteen appointees were from an engineering background, and all others were from the natural sciences. Though two were biologists, there were no ecologists. Furthermore, there were no social scientists at all

— no economists capable of addressing cost-benefit analyses, no public policy analysts, no sociologists to address issues of discontinuity in social structure. And there was no one trained in philosophy or linguistics to enable them to analyze the discourse and paradigms used to promote nuclear amnesia in this controversial area.

Other more institutional biases entered their process. The Regina open house had an AECL and Ontario Hydro display in the same room with the FEARO display. I suppose you could argue that the AECL was informing the public about its nuclear waste storage proposal rather than promoting it, or that Ontario Hydro was informing the public about how nuclear wastes are produced by its Candu reactors, rather than promoting them. However, the presence of these displays gave no assurance that the FEARO was at arm's length from the nuclear industry. If the FEARO stayed true to its mandate to exclude issues of energy policy, what on earth was Ontario Hydro, Canada's main operator of nuclear power plants, doing there? To have any semblance of balance, there should have at least been a fourth display, one on alternative energy technologies like wind and solar, which don't produce nuclear wastes.

Those travelling with the FEARO's open house expressed concern at the sparse public interest in its Regina information session. It's noteworthy that just before the FEARO's Panel came to town, the AECL held a meeting and press conference on its nuclear reactor waste proposal, which stole any of the thunder that might have come with the hearings. It's not surprising that many of the public see the nuclear industry, regulators and reviewers as parts of a unified whole.

The reason for holding these preliminary sessions, and for intervener funding, was supposedly to create an inclusive set of questions, from all angles, prior to the AECL doing its EIS. Such broad input at the start could be seen to counteract the more systemic biases in the process, and ideally the Panel would reflect this exchange of perspectives and questions in its reports. Institutional and ideological bias, however, could easily undermine this stated objective. The money to assist organizations to prepare for hearings came from the proponent, the AECL. (I suppose a case can be made that its distribution through a FEARO committee added some fairness to the process.) And a distinct bias favouring the AECL proposal was reflected in the questions posed in "Socio-Economic Issues Must Be Considered," in Volume 1, Issue 2 of the 1990 FEARO publication, *Dialogue*. These questions reveal a clear structure and orientation. They start by assuming economic growth and local benefits, then raise the fiscal and social problems that may result, but conclude with an emphasis on mitigating these repercussions. The sequence of questions can be seen as an invitation to the AECL to fill in the blanks with convincing arguments about economic prosperity, public participation and social mitigation — without the need to even consider site-specific constraints or the deeper ecological and energy policy questions.

There were no questions posed about the capital costs of such job creation or the loss of economic benefits to other communities due to capital-intensity. Nor were there any questions about the effects of anxiety regarding nuclear wastes over many generations on the morale and quality of life of the region. Nor whether communities facing a depressed economy would even consider such a facility if alternative energy and development paths were envisaged and resourced. This was vital because at the time the Meadow Lake and District Chiefs were being "wined and dined" about the possibility of having the AECL's deep rock nuclear waste project in their area in north-western Saskatchewan, and the U.S. government was also targeting impoverished Indigenous communities for nuclear waste storage. Nor was there consideration as to whether the creation of such a facility would make it more likely that other countries, like France, who are dependent upon Canadian uranium, would want to send their nuclear reactor wastes here for "disposal" (i.e., storage).

The Meadow Lake and District Chiefs had called for a moratorium on uranium mining back in 1977, and, at the time of the FEARO Panel's deliberations on nuclear waste storage, an Indigenous grandmother-inspired activist group opposed the local economy becoming complicit in the radioactive contamination of Mother Earth. Anyway the economic prospects from nuclear waste storage weren't compelling. Using estimates from earlier AECL reports that there would be 2,500 person-years of employment resulting over the fifty years needed to construct deep rock storage, and taking what would likely prove to be a severe underestimation of the costs at $7 billion, we end up with a figure of $2.8 million per full-time job. This is obviously not a cost-effective way to create employment, except, perhaps for those in the AECL already facing unemployment due to the lack of demand for their toxic products. Wouldn't it make more sense to put this kind of investment into alternative economic development and technologies that don't produce wastes that threaten eco-systems for many millennia?

Environmental Review Part of Nuclear Expansion Strategy

While the IUC hoped this FEARO review would be a step towards a more ecologically sane approach to energy policy and technology, it was more realistic to see its process as interlocked with the nuclear industry's strategy for expansion. This is not to question the motives of the members of this particular panel, but only to point out that they were acting from a very skewed mandate in a highly biased institutional environment. The nuclear industry was trying to counteract a serious credibility problem; the CNA's Business Plan reports that in November 1988 the CNA's own Decima polling showed that two-thirds of Canadians didn't think the nuclear industry "is capable of handling its waste." The CNA's strategic documents, which led to its multi-million dollar advertising blitz, showed it realized this issue was its major "public relations" obstacle.

The futures of the AECL, Ontario Hydro and now also Cameco, the three largest corporations in the Canadian nuclear industry, depend upon overcoming the huge credibility gap over the buildup of nuclear wastes. So what better plan than for the AECL to go across the country to have a general review of a highly general proposal for nuclear waste management, and hopefully, to have the Panel say it is a reasonable route to follow? (Later we'll see that, in 1998, the FEARO Panel's report said something like this, though it did not say such a plan had public acceptance.) Since its review wasn't site-specific, it was not binding, nor did it require specific validation. What better way to appear to have found a solution without having to find one? The public relations pay-off for the industry could be vital if this helped to overcome some skepticism among the Canadian public about the AECL's unforgiving technology.

Nevertheless, the conflict of interest that underlies this industry continued to shine through its maneuvering. The fact that the proponent in the hearings, the AECL, and the Panel itself both reported to the Minister of EMR — a minister who was known to be pro-nuclear and who, interestingly, represented the Manitoba riding where the AECL's Whiteshell research station was located — made the process seem somewhat circular. While the Panel also reported to the federal Environment Department, it was the EMR Ministry that carried the clout in matters of energy policy. But, as we have already seen, the Panel wasn't able to review energy policy. What a set up!

The added fact that the AECB, the nuclear regulatory body at the time, also reported to the Minister of EMR showed just how incestuous the nuclear industry had become. The AECB was actually directly involved in the nuclear industry's attempt to expand into Saskatchewan. Leaked documents from the AECB to the Treasury Board[5] indicated that the AECB knew it had a major part in the promotion of the industry. It came as a surprise to us in Saskatchewan that, in its October 16, 1989, *Submission to Treasury Board*, the AECB had already requested and received fourteen new person-years of staff and over $1 million to license "the Candu-3 and new uranium mines in Saskatchewan," even though these projects had never faced assessments. The AECB document to the Treasury Board had argued that without these new resources, which it received, "The marketability of the Candu-3 may be prejudiced as it relies on up-front licensing to reduce its capital costs to make it competitive."

The regulators in this highly integrated and secretive industry were admitting their role in helping with the expansion of the nuclear industry.

When we heard that the FEARO Panel on nuclear reactor wastes was coming to Saskatchewan, we were rightly puzzled, but perhaps the Panel already knew something we didn't — that a decision to promote the construction of a nuclear waste facility in Saskatchewan had already been made. If nuclear regulators could be that complicit with the industry, it is understandable why Canadians had such a hard time believing that a FEARO review process, hamstrung by a biased and limited mandate, could provide them with the kind of independence required to finally address the problem of nuclear wastes free from nuclear politics.[6]

The AECL's EIS wasn't completed until 1994, and the Panel didn't report until 1998. And over its prolonged sitting it apparently had been listening to the loud cries of skepticism coming from across the country.[7] While it noted that the AECL established some technical credibility at the conceptual level, it concluded the "AECL concept for deep geological disposal has not been demonstrated to have public support.... In its current form it does not have the required level of acceptability to be adopted as Canada's approach for managing nuclear fuel waste."[8] Soon after, in 2000, the discredited nuclear regulatory system went under a necessary face-lift: the AECB was replaced by the Canadian Nuclear Safety Commission (CNSC), and the *Nuclear Safety and Control Act* replaced the *Atomic Energy Control Act*, which had been in place since 1946. However, the federal government was still left with no answers or direction for addressing the accumulating nuclear waste crisis. Consequently in 2002 it set up the Nuclear Waste Management Organization (NWMO), and after three years more of public "stakeholders' dialogue" it recommended what it calls "adaptive phased management" of Canada's nuclear wastes — a euphemism for "we haven't got a plan." This all-industry group recommended Saskatchewan as one site for "permanent" nuclear waste storage, and it used the NWMO as a platform for promoting the insidious AECL-Cameco doctrine that since Saskatchewan benefitted from the sale of uranium the province has to assume its responsibility for the end-uses and take back nuclear waste.

All the NDP governments who had blindly promoted uranium mining for so many decades were getting "payback" karma from their colleagues in the nuclear industry. And after fifteen years more of deliberation there still was no credible and acceptable nuclear waste storage strategy. There had been no real learning curve, for we were back where we began, still being asked to make a leap of faith — and we'd had another decade and a half of amassing even more undisposable nuclear wastes.

Notes

1. Walter Robbins, *Getting the Shaft: The Radioactive Waste Controversy in Manitoba*, Queenston House, 1984.

2. Bill Adamson, *Climate Change and Nuclear Power*, ICUC Educational Cooperative, 2006, available at <http://www.icucec.org/climatechange.html> accessed June 2007.
3. Steve Tetreault, "Yucca Mountain: Nuclear industry makes offer, State would get millions for temporary storage," Stephens Washington Bureau, Sept. 21, 2006, available at <http://www.reviewjournal.com/lvrj_home/2006/Sep-21-Thu-2006/news/9783271.html> accessed June 2007.
4. *Dialogue*, Vol. 1, Issue 2, p. 2, Ottawa: Federal Environmental Assessment Review Office (1990).
5. "AECB attacks its record in quest for money, staff," *Globe and Mail*, May 28, 1990, p. B1–3.
6. A good overall source on the politics of nuclear wastes remains Anne Wieser (ed.), *Challenges to Nuclear Wastes: Proceedings of the Nuclear Waste Conference*, Winnipeg, September 12–14, 1986. In this I discuss the importance of including uranium tailings in any discussion of nuclear wastes. Also see <ccnr.org> for updated information.
7. *Report of the Nuclear Fuel Waste Management and Disposal Concept*, Environmental Review Panel, FEARO, Ottawa, 1998.
8. The report is summarized and analyzed by the Canadian Coalition for Nuclear Responsibility (CCNR). See links at <www.ccnr.org>.

chapter 11

ENTERING THE NUCLEAR DEN

By the early 1990s the Canadian nuclear industry was counting on Saskatchewan as a base for legitimacy and expansion. And momentum seemed to be with it. Uranium mining continued to expand. A private consortium was floating the idea of building a small Candu reactor, and the north was being considered as a site for nuclear reactor wastes. It was no accident that the CNA decided to hold its annual meeting in Saskatoon on June 10–12, 1991.

Suffering from a deserved image of being a closed system, the industry tried at this meeting to appear more open to public concerns. I was invited to sit on a panel discussing the prospects of building a Candu-3 in Saskatchewan, but it turned out I was the only critic on the panel, the token "anti-nuke." Worried that my participation might lend legitimacy to the meeting's sponsors, I consulting with several non-nuclear groups before deciding to attend. Initially, I was listed as an "anti-nuclear advocate," but I told the organizers that, to be fair, other panel members should be listed as "pro-nuclear advocates." In the final program we were all designated by our institutional titles only.

While I spoke inside, non-nukes demonstrated outside; so I thought of it as a "double-whammy." On this sunny summer day I definitely felt like we were making gains in our understanding and influence — why else would the CNA feel the need to invite one of us? But I knew I must be thoroughly prepared, as my assertions were not going to be accepted lightly. I spoke to a full house of Canada's "nuclear establishment" and to my astonishment no one directly challenged my facts. After my talk, several delegates, some of whom were technological suppliers to the AECL and Ontario Hydro, privately expressed sympathy with my position, some indicating they would prefer to be suppliers to renewable energy systems.

The other four panelists, including the chair, were nuclear proponents or sympathizers, so I provided some semblance of balance. My message no doubt would have been better received at a solar or wind energy conference, but as an educator I saw little point in speaking only to the converted. I encouraged those present to momentarily disengage from their professional and institutional stakes in the nuclear industry, to allow some critical reflection on the industry's ideology and to open their minds to alternatives.

The panel addressed the question of whether a Candu reactor should be part of Saskatchewan's future. The question was a bit rhetorical. The fact that the local population was being bombarded by costly pro-nuclear CNA ads, that networks of pro-nuclear personnel were being consolidated within the government, universities and media, and that one-sided promotions pervaded our public schools, provided compelling evidence of a concerted plan to ensure that the answer would be "yes." The president of the CNA was quite candid about the commercial motives for being in Saskatoon, saying, "There is growing interest and support in Saskatchewan for the construction of a demonstration Candu-3.... It is therefore no coincidence that this year's conference is being held in Saskatoon...."[1]

The CNA apparently believed its own propaganda and was trying to live a self-fulfilling prophecy. However, though their ads attempted to associate nuclear reactors with mothers who care about health, children who care about the environment and parents and grandparents who care about children, the controversy about nuclear energy would inevitably return to the public policy arena. It was to this complex public policy history, which can't be grasped in manipulative promotions, that I turned. I began by looking at the history of Candu exports.

Looking Candu History Square in the Eye

Candu export history is not something about which informed ethical Canadians can be proud. Nor has the Candu export history been economically responsible. Small research reactors were initially sold to India and South Korea, and, as we should all know, the Indian plant produced the plutonium that was used to explode India's first atomic bomb, in 1974.[2] Even though India's motives were already questionable, a commercial Candu, completely financed by Canada, was built there in 1972, and another was built in Pakistan that same year. Concern that Pakistan was interested in using its Candu to build the first Islamic bomb grew after a BBC report on this question in the late 1970s. Canada refused to sell Pakistan any further fuel or parts for fear of this happening, but it was too late and Pakistan went on to become a nuclear weapons power.

India wanted the bomb because China had it, and Pakistan wanted it because India had it, and Canada's Candu salesforce were willing partners to this arms race. Business dealings over a Candu sale to Argentina in 1973, when it was still a repressive military dictatorship, add to this sorry story. The president of the AECL admitted to the federal Public Accounts Committee that he had been paid consulting fees by a subsidiary of the AECL's Italian partner. Total losses to the Canadian taxpayer on this deal were reported in the January 28, 1977, *Globe and Mail* to be $130 million, as cited in the AECL's 1977 annual report. In November 1981, an assistant to the Argentinean president commented that Argentina's energy problems could be solved much more cheaply without nuclear power and that the main rationale for the country's nuclear program was not economic, but military and strategic.[3] So,

Canada's Candu promoters almost played a role in fuelling a South American arms race between Argentina and its rival nation at the time, Brazil.

The next country to be targeted was South Korea, yet another country with an authoritarian regime at the time. Desperate for a sale, the AECL promised $300 million in credit, three-quarters of the total cost. A $17 million agent's fee was also paid, about which the Auditor General said there was insufficient documentation for over half. This led to speculation about the use of bribes. But we learned nothing from this fiasco; the very next deal — the biggest ever for the Candu industry — was with the notorious Ceausescu regime. In 1980 Canada granted a $1 billion line of credit to Romania to finance the sale of five Candus. Revelations since Ceausescu's overthrow further discredited the Candu industry. After inspecting the shoddy construction and falsified records in 1990, the International Atomic Energy Agency (IAEA), which operates under the United Nations, recommended no further work without major corrections. Repairs were estimated at $40 million. The AECL requested an additional $250 to $350 million from the federal government to complete the five Candus, but the public outcry was immediate, and the *Financial Post* carried columns and stories opposing further loans to Romania.[4]

Romania was one of the most authoritarian and economically backward countries in Eastern Europe. Ceausescu's grandiose pro-nuclear scheme was to build nineteen reactors by 2001, and food was withheld from domestic consumption to raise cash from exports for financing this megalomania. Nuclear debts were repaid at the expense of the wellbeing and human rights of the people. Electricity was to be generated by 1985, but this never occurred. In supporting the additional multi-million dollar loan, the president of Candu Operations predicted the plants would be operating at the earliest by 1994, an incredible fourteen years to come on-stream, but they weren't functional until 1996. Little wonder that other countries have not been scrambling to make a deal.

An even more frightening scenario might have unfolded in 1974, when the AECL was among the nuclear companies trying to make a reactor deal with Iraq. Yves Girard, advisor on nuclear affairs for France's Department of Energy, is quoted at length in the book, *The Islamic Bomb*, about these dealings: "The nuclear merchants wanted to get a piece of the nuclear action. It was within the realm of salesmanship to point out the virtues of the product, including its ability to produce plutonium." And, according to the authors, Girard

> was especially critical of the surprisingly competitive Canadians for doing so. They made a number of trips to Baghdad where they sat around their hotel puffing the virtues of the Candu natural uranium

> reactor to their potential clients, hinting broadly at its excellence in producing the deadly substance and even more broadly at the possibility of keeping safeguards to a minimum.

But there is more:

> What bothered Girard most about them was not that they had tried to sell the Candu that way. It was their later hypocrisy in pointing to the French sale as a danger for nuclear proliferation when actually they had desperately wanted the sale for themselves and had indicated no concern whatsoever whether Iraq got the bomb or not.[5]

In an increasingly competitive world market, Candu officials know better than anyone how disastrous their export dealings have been. In its January 1990 *World Status Report* on nuclear power, the *Energy Economist* noted:

> The Canadian nuclear industry has made two basic mistakes; to be born in a country with almost unlimited reserves of cheap hydropower; and to make some fairly unfortunate choices, namely India and Romania, in its search for export customers. At present, Canada needs few, if any, new power stations, and the export market no longer exists.

On April 11, 1981, *The Globe and Mail* reported that Prime Minister Trudeau was asking, "Should we get out of the business?… One can't hide that it is a very costly program to maintain. Unless we are successful in selling many more Candus, then it will be a more difficult decision to take."[6]

No export sales were made through the 1980s, a time when one reactor a year was considered essential for the industry to pay its own way. It took a full decade to make just one sale, the 1990 deal with South Korea. In its August 1988 report, *Nuclear Energy: Unmasking the Mystery*, the Standing Committee on Energy, Mines and Resources (EMR) was quite candid about the predicament of the AECL and its Candu sales staff, stating, "It is clear that reactor sales alone will not carry the AECL through the coming low period without financial assistance. To minimize the need for federal funding, the AECL must look to other business opportunities." It continued,

> One such opportunity is the Canadian Submarine Acquisition Program.… A task force of senior AECL Research and Candu Operations staff has recently been acting in an advisory capacity for the submarine program.… It has been suggested that AECL may be named the prime contractor for all nuclear elements of the submarine program.

This Standing Committee "unmasked the mystery" of the Canadian nuclear industry, confirming that the Canadian public didn't need nuclear power, and new markets were needed for the AECL to survive. Thankfully the nuclear subs were not built. However, without this contract, the AECL's dependency upon federal grants increased and it was even more desperate to find new markets. This is the context within which the push to build a Candu-3 in Saskatchewan must be understood.

In its first forty years of operation, the AECL sold only eight reactors to four countries, most in its disastrous Romanian deal, and after a half-century the total sales were only eleven to five countries. During this period, according to the Economic Council of Canada, the industry received $12 billion (in 1981 dollars) in government subsidies, which are never going to be returned.[6] After OPEC increased oil prices in 1977, there was growing pressure for energy conservation and energy alternatives. However, from its inside track, the AECL was able to "corner the market" on government subsidies. By 1979, according to the Parliamentary Sub-Committee report, *Alternatives to Oil*, even with the growing environmental movement and pressure for renewable sustainable energy, 64% of all federal research and development grants still went towards conventional nuclear power, compared to only 21% for all renewable energy sources and conservation combined.

The Mulroney Tories came to power in 1984 promising a review of federal support for nuclear energy; however, once in power, they followed the same path as neo-conservative governments in Britain and the United States and began phasing out grants for conservation and renewable forms of energy. Federal support for the AECL doubled from $100 to $200 million a year from 1984 to 1989, and it wasn't until the federal Green Plan was introduced in December 1990 that grants for renewable energy were reintroduced, but these were nowhere near the amount still going to the nuclear and oil industries.

How this reversal of policy occurred is most revealing. Due to the industry's sorry showing and the Tory policy of fiscal restraint and privatization, the decision was made in 1985 that AECL grants were to be reduced from $200 to $100 million a year over five years. This was to lead to the selling off of the Candu industry. Major job cuts and layoffs occurred within the subsidy-dependent AECL, and an estimated 200 of a remaining 1,000 jobs were to be cut by the end of 1989. Then in 1989 the new minister, Jake Epp, came to the rescue, announcing restructuring of the AECL so that it could enter into joint ventures with private firms. In view of the British failure to privatize the nuclear industry, the Canadian government was searching for other ways to cut its losses, which included food irradiation, shrunken Candus (the proposed Candu-3), and enlarged Slowpokes.

Rather than reducing federal grants to the AECL, as previously stated, Epp announced that support would now rise to over $200 million annually for seven years. At the time the AECL was still receiving $140 million in federal grants, but officials were pleading that further cuts would jeopardize hundreds of jobs at AECL plants. The fact that vulnerable AECL jobs at the Whiteshell Nuclear Research Station in Manitoba were within Epp's federal riding of Provencher made this rescue suspect.

Marketing the Candu in Saskatchewan

Well before the 1989 proposal to build a private Candu-3 in Saskatchewan, the AECL was developing a new strategy to market its hardware. The design of the 450 MW Candu-3 attempted to overcome past failures in the export policy by reducing the long period of construction through a new, modular system and by having a less costly smaller reactor that could be marketed in more varied circumstances and that could compete with small-scale coal-fired plants. The AECL's authors said,

> The Candu-300 [commonly known as Candu-3] is an economical small nuclear power station, ideally suited to countries and utilities with uncertain load growth, small grid size and/or limited financial resources. It makes the nuclear power option available to a very broad spectrum of countries and utilities that have previously been effectively excluded.[7]

There is nothing in this AECL paper about the Candu-3 being "environmentally friendly." Rather it contained only an engineering and marketing argument. All the hype about nuclear power being the panacea for acid rain and global warming came later, as part of a clever CNA promotions campaign. But the AECL's technical paper did address the matter of speeding up licensing to meet regulations. It refers to "a close dialogue" with the AECB over what it called "up front" licensing of the Candu-3. The close relationship between the AECL, which markets the Candu, and the AECB, which regulates it, had long created a credibility gap for the government-run and regulated industry. That both organizations existed within the same ministry (the EMR) and that their appointees always seem to be pro-nuclear compounded this problem. The appointment — then de-appointment — of non-nuclear scientist Dr. Ursula Franklin of the University of Toronto to the AECB in 1985 is a case in point.

We finally confirmed from leaked AECB requests to the Treasury Board that the regulatory agency was part and parcel of the AECL's marketing strategy for the proposed Candu-3. In its October 16, 1989, application the AECB sounded more like a nuclear proponent than a regulator,[9] making a pitch for extra staff to "correct recognized deficiencies in the effectiveness and openness of the nuclear regulatory process." This was an admission of the shortcomings earmarked by environmentalists, but predictably denied by the AECB in the media. One deficiency that it highlighted was "to remedy the most urgent safety and openness issues." In view of all the public reassurances of the greater safety of the Candu, since the 1986 Chernobyl accident, it is revealing that the AECB submission stated that "Candu plants cannot be said to be either more or less safe than other types."

The other urgent issue was "the licensing of Candu-3 and new uranium mines in Saskatchewan." One of the arguments for additional resources was that "the marketability of the Candu-3 may be prejudiced as it relies on 'up front' licensing to reduce its capital costs to make it competitive." It was astonishing to see requests for AECB staff to license a nuclear reactor that hadn't been formally proposed and wasn't involved in any environmental review process. The cynical premeditated way the nuclear industry was working hand-in-glove with regulators to control government decisions was also demonstrated when the submission noted: "Between 1988 and 1993 four new underground mines are expected to come into full operation." At the time, environmental reviews and public hearings didn't exist for these mines — yet the AECB was granted fourteen new person-years of staff and additional funding of $1.3 million for licensing the Candu-3 and proposed uranium mines. (After the uranium mine spill at Wollaston Lake in 1989 public pressure finally led to a FEARO panel for these new mines.)

It's apparent from the leaked document that the AECB saw itself as an integral part of the AECL's new marketing strategy. It's also clear that the AECB saw itself as part of the process of managing the crisis of legitimacy facing the nuclear industry, for it unabashedly stated: "Gaining public acceptance and confidence that governments can safely deal with radioactive wastes appears to be of equal or greater difficulty than solving the technical problem." The AECB was clearly on the same wavelength as the AECL (and later the NWMO), which believed that the waste issue must be managed as an issue of perception not long-lived ecological toxicity.

While the AECB was arranging to be the "regulatory" part of the AECL's Candu-3 marketing strategy, the CNA was mounting its parallel public "education" campaign. The corporate media was a willing partner, and pro-nuclear messages flooded the Canadian airwaves. The *Financial Post* carried a two-page feature, "A Report on Canada's Nuclear Industry," on June 13, 1988. The small print at the top of this "Report" stated: "all editorial material supplied by the CNA." Much erroneous information was in this supplement, including a "story" (i.e., advertisement) headed, "Native Workers Bring Stability to Saskatchewan's Uranium Mines." What started as 400 "Native employees" became, by the end of the "story," 400 "northern/Native residents." This flies in the face of the fact that Indigenous people in the north had not received the jobs they were promised when the expansion of uranium mining was given the go-ahead in the late 1970s.

Recall that the 1987–88 CNA Business Plan discussed in chapter 5 was to "develop a media briefing book to be used as a reference document by media and which can be provided to the media through editorial board meetings, and media briefings and visits." A major goal was to "maximize opportunities for positive third party op-ed articles." The desired results included, "Have industry viewpoints considered in coverage of material related to the industry… [and] Have four positive op-ed articles published annually."

Based on the pro-nuclear coverage occurring in Saskatchewan's corporate media since 1987–88, these CNA's objectives were met. I have clipped "news" stories and letters to the editor in the Regina *Leader Post* since the first Candu-3 project was publicly announced in late 1988. Pro-nuclear views dominated headlines, received many more column inches and were reflected in more and longer commentaries. There also was a twenty-four-page supplement, "Saskatchewan's Energy Alternatives," in the April 29, 1989, La Ronge *The Northerner*, which was filled with pro-Candu-3 promotions. A four-part opinion-commentary in the Saskatoon *Star Phoenix* from July 6–13, 1989, by columnist Paul Jackson, promoted the Candu. His July 6 column noted, without concern, that AECL and CNA officials accompanied Colin Hindle as he promoted his private Candu-3 proposal.

The CNA was engaged in big league marketing — with public money. The AECL, one of the CNA's major fee-paying members, received millions of dollars from the government and taxpayers. In the communist Soviet Union, most notably during and after the Chernobyl accident, the nuclear industry controlled information through the government. In Canada, a democracy, the industry learns the language of review and regulatory processes without actually entering the spirit of them, and in collusion the corporate media distributes its own "news" promotion.

A Closer Look at the Environmental Ticket

The strategy for marketing the Candu-3 relied heavily on promoting it as the environmental friendly energy choice. In its 1990 World Status Report on nuclear power the *Energy Economist* said that it was "precisely the environmental ticket, unpredictable as it is, which nuclear power needs to play." The "greening" of the nuclear industry, especially since the 1986 Chernobyl reactor explosion and fire, amounts to new packaging to better sell its product, and the attempt to present nuclear power as the environmental answer to acid rain and global warming is particularly flawed. As explained in chapter 9, nuclear plants replacing coal-fired plants would barely make a difference in greenhouse gas emissions since only one-sixth of these gases come from fossil fuels used to generate electricity and coal is only one of these fossil fuels. Transportation is a growing cause of CO_2 from fossil fuels, and the destruction of the rain forests also contributes to global warming by reducing their absorption of carbon from the atmosphere. Furthermore, the nuclear industry is actually a very heavy user of the very fossil fuels that contribute to acid rain and global warming,[8] and the industry is even suggesting that the Candu could help produce Alberta's filthy heavy oil.

Sometimes the environmentalism of nuclear proponents gets downright absurd. The attempt to use the support for nuclear power of Gaian theorist James Lovelock against environmentalists opposing a Candu-3 for Saskatchewan is one example.[9] In making this case in the April 26, 1991, Saskatoon *Star Phoenix*, University of Saskatchewan nuclear physicist Dr. Kupsch failed to mention that Lovelock is less interested in humanity than in Gaia *per se* and that he doesn't have a good track record when it comes to predicting specific environmental catastrophes such as the thinning of the ozone layer or the Chernobyl disaster.

There has also been an erroneous attempt to get the nuclear industry on the bandwagon of "sustainable development," but as we have shown, the Brundtland report recommends renewable energy, not nuclear energy, as fundamental to this quest.[12] Many other sources saw the emergence of the renewable energy sector as a viable and necessary option, and there is good logic supporting this optimism. For example, if lighting accounts for a quarter of U.S. electricity, and if, as the Rocky Mountain Institute suggests, as much as 92% of this can be saved by efficient lighting and daylight-saving strategies, then total electricity use could be reduced to 77% by this act alone.[10] Early on in this debate, a 1990 study by the Union of Concerned Scientists, "*Cool Energy: The Renewable Solution to Global Warming*," concluded that "as much as 50% of U.S. energy could be provided by renewable energy sources by 2020, assuming no overall growth in energy consumption."[11]

Around the time of the CNA meeting in Saskatoon there were almost daily reports in the media about progress in the renewable energy sector. For example, a story in the April 4, 1991, *Globe and Mail*, "Finding could halve cost of solar power," reported on a method to use inexpensive, low-purity silicon to convert sunlight into electricity. Worldwatch Institute's *World Resources: A Guide to the Global Environment*, published in 1990, pointed out that in some countries the renewables were already supplying more primary energy than was produced by nuclear power. It lists Canada as one example.

Sweden's Phase-out of the Nuclear Phase-out

The contest between nuclear power and non-nuclear alternatives has been particularly heated in Sweden. On the surface, the constraints on introducing non-nuclear energy there seem immense. In the 1980s, Sweden's twelve reactors generated 68 terawatt-hours (TWh), or 68 billion kilowatt-hours (KWh) of electricity, almost half the country's total. In the early 1990s Sweden had the largest per capita consumption of nuclear power in the world. In 1988 Sweden was the first European nation to agree to hold CO_2 greenhouse gases at 1988 levels, 70% of the potential hydro-electricity had already been developed, and the four remaining remote rivers would "only" be able to generate 15–20 TWh of electricity. Skepticism about Sweden's transition to a non-nuclear society was shown in the *Energy Economist*'s 1990 World Status Report on nuclear power:

> Neither the public nor anyone else has been able to explain how a country deriving half its electricity from nuclear power; a country which is the highest per capita user in Europe; a country which wants to build no new hydro plants and has banned any increase in carbon dioxide emissions, let alone sulfurous gases; how such a country can take the lead in the great escape from nuclear.

With all these constraints the Canadian and international nuclear industry looked with trepidation on Sweden's 1980 referendum to phase out nuclear energy. After the Three Mile Island nuclear accident in 1979, the ruling Swedish social democrats reversed their position and supported such a referendum. They however manipulated the wording to try to neutralize the momentum for a total phase-out. Instead of two choices, the social democrats added a third, which split support for a "quick stop." Support for no more nuclear power plants and the phasing-out of the remaining six plants was 38.7%; whereas only 18.9% supported completing the six reactors under construction and only closing them down as electrical demand and employment opportunities allowed. However the new, third option, stressing public ownership of nuclear power and some energy conservation, in addition to completing the six reactors, got 39.1%. By combining the two "slow stop" options, the "quick stop" was said to be defeated, and in 1980 the Swedish parliament decided to limit reactors to twelve and do the phase-out by 2010. It also interpreted the results to support an unproven nuclear waste storage system.[12]

Before politics "took over," Swedish research on three non-nuclear, non-hydro sources of electricity (conservation through energy efficiency, the use of biomass, and the use of wind power) already looked very promising.[13] A Lund University study indicated that shifting to energy efficient lighting would save 80% of the electricity used for lighting, reducing electrical consumption from 140 TWh in 1990 to 96 TWh by 2010. Another study suggested that with no net increase in greenhouse gases, biomass could produce 20–35 TWh after ten years. Finally, a government study suggested that four thousand wind generators, with only one-third built on land, could replace up to half of the electricity produced by Sweden's twelve nuclear plants. With just these three strategies, the need for about 44 TWh could be reduced and another 54–69 TWh produced. With a shift of capital, labour and policy towards these technologies, it would be possible to replace the 68 TWh produced by nuclear reactors.

If Sweden could do it, probably any country could. The industry strenuously organized against this scenario and was active in the 1990 election, which saw the defeat of the social democratic party, which had orchestrated the "slow stop" option. The new conservative government tried to cancel the phase-out, but in the wake of mass protests abandoned the idea.

Italy had already announced a nuclear phase-out in 1987, Belgium did so in 1999 and Germany in 2000. And Sweden went on to shut down nuclear reactors in 1997 and again in 2001. However in 2003, under pressure to maintain a competitive nuclear industry within Europe, a social democratic government held another referendum, freeing it from the 2010 deadline for shutting down the remaining reactors. In 2006, the centre party, which had previously supported the nuclear phase-out, also changed its policy.

That the nuclear phase-out is not going to happen by 2010 is self-fulfilling. The social democrats, who have been in power most of the time since the 1980 referendum, did not curtail energy consumption through aggressive conservation and renewable energy programs to make the phase-out possible. Now, the Swedish coalition conservative government has announced it won't decommission any more reactors, instead supporting plant-life extension, similar to what Ontario Hydro has done with the Bruce reactors.

How similar this all sounds to how the Saskatchewan NDP has mishandled energy policy. Narrow economic interests rather than broad energy or environmental policy considerations have been paramount, demonstrating how the struggle to create a non-nuclear society may involve taking two steps forward and one back. The economic structure itself needs reforming. "Jobs at any cost" is not sustainable. And greed and power has to be brought under control, and Mother Earth, creation, evolution — whatever you wish to name "life" — must have more value than profits.

Exposing the Saskatchewan Candu Caper

What did all this mean for the WPDA proposal to build a private Candu-3 in Saskatchewan? Remember, this proposal appeared shortly after the Devine government announced its plan to privatize some of the province's utilities. Prior to the 1988 Saskatchewan election, jobs at the Regina Oil Up-grader helped Devine win a second term, and it appears he hoped a fast-track Candu-3 would play a similar role in the buildup to the 1991 election. Opposition to this energy path, however, was widespread, and Devine and his colleagues were about to pass into political history. They were replaced by the Romano-led NDP, which was not only against a Candu-3 in the province but a large majority of the membership and party policy supported the phase-out of uranium mining. After the NDP won, its November 1991 convention passed a resolution by a four to one margin among the thousand delegates calling for the scrapping of the SaskPower-AECL agreement to do research on a Candu-3.[14]

The Devine government apparently took some of its ill-conceived coaching from advisors to British Prime Minister Margaret Thatcher. Under her government the Central Electricity Generating Board (CEGB) tried to privatize its eleven nuclear reactors in England and Wales, with the taxpayers ready to "sweeten the deal" by providing over a billion pounds towards decommissioning, fuel processing and fuel disposal costs. The same arguments used in Canada (that nuclear power diversifies energy supply and doesn't cause acid rain or global warming) were used to try to sell the British utility — but to no avail. All along the CEGB had emphatically predicted the need for another 15 GW of capacity by the year 2000, to add to the country's 55 GW capacity. Meanwhile, according to the article, "Electricity privatization: Saving grace," in the December 10, 1988, *The Economist*, 6.5 to 12.5 GW capacity could actually be saved by the year 2000 simply by using available energy-saving technology. Combined heat-and-power, or co-generating, schemes could save the need for up to another 10 GW capacity. Further, the cost of following this route would be one-sixth to one-twelfth that of building extra capacity through nuclear reactors.

Governments who helped create the fiscal crisis may still want to believe their own miscalculations about the need for nuclear power, but the business world is rightly more skeptical. The 1990 *World Status Report* on nuclear power predicted that the rising costs of decommissioning will be the Achilles' Heel of nuclear power economics. Were the taxpayers to be more fully informed of these hidden costs, their opposition to nuclear power would most certainly increase.

During those days of intense conflict over the Candu-3 proposal, we also heard a lot about the immense costs of servicing the national and provincial debts and were told to tighten our belts. Yet we heard nothing about how much of this debt came from electrical utilities bent on expansion and trapped in a cycle of energy growth to increase sales to pay the debt on past capital-costly mega projects. Ontario Hydro, the AECL's largest consumer of reactors, had a $12 billion debt, about the same size as federal government subsidies to the nuclear industry at that time. SaskPower's debt was over $2 billion, and though 1988 was considered a successful year, with a $122 million operating profit, the debt cost the utility $222 million to finance.[15]

Nevertheless, with the help and backing of the AECL and CNA and the blessing of the Devine-appointed SaskPower management, the WPDA continued to promote a private reactor. Reactor sales reps came in and out of the province, treating it as a lucrative Third World market. When the costs of nuclear power were seriously discussed, including such relevant factors as the million-dollar cost of each kilometre of electrical transmission lines, for example, WPDA president Colin Hindle tried to justify a Candu-3 as a means to meet the electrical demand of a new heavy oil upgrader on the Saskatchewan-Alberta border.[16] Another clever idea to save the AECL, which persists to today, was born. This idea of using a reactor for heavy oil was again floated, in 2003, by the Alberta government and more recently by a past Saskatchewan NDP cabinet minister, Dwayne Lingenfelter, who now works in the Alberta oil industry. This would have nuclear energy contributing directly to fossil-fuel production and global warming, which shows how little the environmental ticket really meant when push came to shove.

The attempt to sell a Candu-3 in Saskatchewan smacks of the same kind of bribery and misinformation typical of AECL's Candu export sales. The employment promised in northern Saskatchewan by the WPDA turned out to involve a cost of $600,000 of investment per twenty-year job. Though posing as a private interest, the WPDA wanted the same federal financial guarantees given to construct first reactors in other places, like New Brunswick, or to help make sales to places like Romania. As with foreign sales, local elites — including within business and the universities — that might glean short-term benefits from a Candu were brought on side. And these promotional "bribes," as in export sales, were mostly done behind the scenes, as with the WPDA's negotiations with economic development boards in Saskatchewan's north; or where only one side of the debate was allowed, as when the Saskatchewan Chamber of Commerce endorsed the Candu-3 after hearing a highly biased, one-sided AECL speaker at its 1991 annual meeting. This practice of secret promotions helping to keep deadly secrets continues to this day.

The desperation of the Canadian nuclear industry was, however, coming full circle. What started as an attempt to cash in on the oil crisis in the 1970s became an attempt to cash in on the environmental crisis in the 1990s. However, environmental groups unanimously favour a non-nuclear path and are too knowledgeable to be fooled by glamorous promotions and grandiose predictions, leaving the nuclear industry with one final desperate argument — economic development and jobs — which was wearing thin, even with the general public. In the late 1970s Saskatchewan was touted as destined to become a "have province" thanks to an impending uranium boom. But by the time of the CNA annual meeting in 1991, uranium revenues to Saskatchewan had only reached about 5% of the high and 15% of the low projections.

This time the gold rush mentality was being perpetuated by a government-dependent reactor industry in deep trouble. When things got rougher for the private Candu proposal, the chair of the AECL Board, Robert Ferchat, said Saskatchewan people shouldn't sell themselves short, "Rather than sell uranium for $10 a pound, why not lease it out and bring it back after its been used in a reactor? Why not put the uranium back where it came from?" he asks.[17] Ferchat apparently promised the province "$100 a pound, 10 times today's market value" for doing this. He didn't mention, or perhaps even know, that according to previous "gold rush" predictions, Saskatchewan was to be receiving $40–50 for a pound of uranium by the late 1980s, without even having to take the AECL's nuclear wastes. (This price range wasn't actually achieved until the late 1990s.) We weren't as gullible as the AECL head apparently thought, and the nuclear merchants would soon have to start looking for yet another place to peddle their costly and toxic wares.

Notes

1. J.M. Reid, "Nuclear Canada: Year in Review," *Nuclear Canada Yearbook 1991:* Annual Review and Buyer's Guide, Canadian Nuclear Association, 1991, pp. 5–9.
2. "Candu: A Review of Performance, Cost & Safety," Glendon Energy Series, No. 1, 1981. Also see G. Edwards, "Fueling the Arms Race: Canada's Nuclear Trade," *Ploughshares Monitor*, undated, pp. 16–21.
3. M. Usher et al., "The Role of Nuclear Power," taped panel discussion. Referenced in Gordon Edwards, "Cost Disadvantage of Expanding Nuclear Power," *The Canadian Business Review*, Vol. 9, no. 1, p. 29.
4. See *The Financial Post*, Dec. 3, 1990. Also see G. York "Romania likely to get loan for reactor, official says," *Globe and Mail*, Nov. 26, 1990.
5. S. Weissman and H. Krosner, *The Islamic Bomb*, New York: Times Books, 1991, p. 91.
6. R. Sheppard, "Candu drought may end in 89," *Globe and Mail*, Jan. 16, 1989, p. B1–2.
7. G.L. Brooks and R.S. Hart, *The Candu-300 Reactor System*, AECL, undated, pp. 11–12.
8. See N. Mortimer, "Nuclear Power and Carbon Dioxide: The Industry's New Propaganda," *The Ecologist*, Vol. 21, No. 3, May/June, 1991, pp. 129–32. The most comprehensive discussion of this is found in Brice Smith, *Insurmountable Risks: The Dangers of Using Nuclear Power to Combat Global Climate Change*. Takoma Park, MD: IEER (Institute for Energy and Environmental Research), 2006. See also Helen Caldicott, *Nuclear Power Is Not the Answer*, New York: The New Press, 2006, pp. 3–18.
9. See summary of this in Jim Harding, "Nuclear Power and Sustainable Development," *Briarpatch*, May 1991, pp. 17–18.
10. Murray Mandryk, "Conservation can alleviate energy woes, experts say," Regina *Leader Post*, Dec. 30, 1988, p. A3.
11. *Cool Energy: The Renewable Solution to Global Warming*, Union of Concerned Scientists, Cambridge, MA, 1990, p. 22.
12. "Nuclear Waste in Sweden: 4 The Referendum," available at <www.folkkampanjen.se/nwchap4.html> accessed June 2007.
13. A. Burke, "Learning to Live without Nuclear Power," *Current Sweden*, No. 372, March 1990.
14. "NDP says no to reactor," Regina *Leader Post*, Jan. 14, 1989, p. D9.
15. G. Brock, "Huge debt overshadows record SaskPower profit." Regina *Leader Post*, Mar. 17, 1989, p. C11.
16. M. Mandryk, "Reactor costs queried," Regina *Leader Post*, Jan. 19, 1989.
17. P. Martin, "Saskatchewan sitting on economic mother lode with uranium," Saskatoon *Star Phoenix*, Apr. 20, 1991, p. F1.

chapter 12

URANIUM–NUCLEAR ALLIANCE

In the fall of 1991 the uranium and nuclear industries launched an integrated expansionist strategy for Saskatchewan. With no Candu sales, the threat of losing huge federal subsidies and growing public pressure to address its long-standing nuclear waste storage buildup, the AECL was "under the gun." Meanwhile Cameco, which was privatized after the merging of the SMDC and Eldorado Nuclear in 1988, had watched uranium prices plummet and the reactor fuel market's failure to expand. In spite of these challenges, and remembering that Saskatchewan had weathered the anti-uranium protests of the late 1970s, there was a possibility that the industry could take advantage of the fact that the province still had the largest operating mines in the world. More Candus would mean more demand for uranium, so why not use the province as a base to launch a new Candu industry? And why not take the nuclear waste back and put it in the "nuclear dumping" ground already being created in the north for uranium tailings? A far-fetched nuclear fantasy perhaps — but it was at the very core of the industry's expansionist strategy.

The non-nuclear movement was cautious about how it responded to the convergence of these issues. Its spokespersons didn't want to sound like they were invoking "a conspiracy theory" as some industry proponents had already dismissively accused non-nuclear "zealots" of being paranoid. Then, in February 1992, I read a report commissioned by the AECL on the "potential impact" of the nuclear industry on the Saskatchewan economy,[1] which confirmed that there was indeed an integrated uranium/nuclear strategy. The scope of its proposal was stunning. The report was finished two days prior to the October 1991 provincial election but not released until February 1992. It may have been timed to coincide with the provincial election as part of a strategy to create momentum for the governing Tories, who embraced nuclear power as part of their electoral campaign. It may have also been timed to put pressure to "go nuclear" if there was a new NDP government. The political connections were all in place. The main author, Roy Lloyd, the NDP government's chief planner in the 1970s, had overseen the expansion of uranium mining and later became president of the Crown's uranium company, the SMDC.

Expanding the Nuclear Fuel System in Saskatchewan

The pro-nuclear bias in the methods and results of this AECL-commissioned report were no surprise, but it is customary to have references in impact studies so that the claims can be double-checked against other information in the field. This "study" had no sources or references, and, as we shall see, all its assumptions were irrationally favourable to the nuclear industry.

The executive summary put an end to all conspiratorial speculation, saying, "From a strong uranium mining base, there is excellent potential for developing other areas of the nuclear fuel cycle in Saskatchewan, particularly enrichment, electrical generation and used fuel disposal." The report advocated Cameco moving into uranium enrichment by 1997 and SaskPower receiving federal aid to build a Candu-3 by the year 2000, with "good potential for constructing up to three 450 MW Candu-3 reactors prior to 2020." It suggested that Saskatchewan be the "repository for disposal of used Canadian nuclear fuel" and leasing "uranium fuel, taking back the used fuel for disposal, which in turn would enhance the market for Saskatchewan."

When the non-nuclear movement opposed the expansion of uranium mining in the 1970s, we warned that this could be a foot in the door for the whole nuclear industry. The attempt to build a uranium refinery in the early 1980s at Warman, near Saskatoon, suggested the industry was serious even then about expanding in the province. But we actually underestimated the plans the nuclear industry had for Saskatchewan.

The AECL's economic strategy of "playing on" the recession affecting Saskatchewan at the time was revealed in an earlier report leaked to environmental groups, a data-base the AECL kept on non-nuclear groups. Referring to the "weaknesses" of the Saskatchewan Environmental Society (SES), page 41 of the January 12, 1989, "AECL Memo, Restricted Commercial" said, "Poor provincial economy has moved environmental issues to low priority which is frustrating group's efforts. Faces strong political will in the province to stimulate economy.... Uranium mining opponents were described as burning out." Later, under the sub-head "AECL opportunities," the report stated, "SES represents a model of how strong, pro-nuclear political will based on economic issues can diffuse or even demoralize opposition."

The 1991 report's executive summary confirmed that the uranium/nuclear industry was embracing a joint, economic stimulus campaign to expand in Saskatchewan: "Uranium mining *has put Saskatchewan in the nuclear industry* and has established a basis of public acceptance to expand into enrichment and nuclear power generation" (emphasis added). This integrated corporate strategy was as political as it was economic and, as we shall see, had little to do with actual energy needs.

The report's methods and logic lacked simple intellectual credibility. The use of broad and vague terms like "prospects" and "potential" weren't consistent with minimum requirements of impact analysis. This "preliminary valuation" would never find its way into the literature on impact analysis. The intent of this report, however, was not impact analysis but the creation of a "golden egg" to encourage the province's maximum involvement in the nuclear fuel system.

Conspicuously late in the report, on page 43, the consultants admitted just how rudimentary was their analysis:

> It should be noted that this input-output analysis provides only a preliminary valuation of the economic impact of the nuclear industry for Saskatchewan. Detailed breakouts on the specific nature of direct expenditures and labour sourcing were not done. In addition, the analysis assumed no structural change in the Saskatchewan economy....

That such a preliminary report received widespread coverage as "newsworthy" is testimony to how the corporate media passively endorses the promotional claims of the nuclear industry. A balanced public presentation of the handling of the issues would have quickly put an end to half-truths and untruths, exposing them as fantasies. At the very least, one would expect a comparison of "risks" and "costs" with "benefits," including analysis of both the environment and the economy. But there wasn't even the pretence of a balanced investigation.

The AECL consultants' report explored opportunities for Saskatchewan to expand into all aspects of the nuclear fuel "cycle." "Cycle" is a misnomer, since uranium tailing waste and nuclear reactor waste remains after milling and fission. "Cycle" gives the false impression that a process is in place that addresses the full spectrum of activities of the uranium/nuclear industry, which is untrue. I therefore refer to it as the nuclear fuel system.

The report discussed conversion (refining), enrichment, fuel fabrication, nuclear power reactors, reprocessing and nuclear waste disposal. It saw "little potential" for conversion (refining) or fuel fabrication facilities in Saskatchewan and "no potential" for reprocessing. (In view of this assessment that there was little potential for uranium refining, it is interesting that this is what the Calvert NDP later targeted for nuclear expansion.) Concentrating on enrichment, the Candu-3 and nuclear waste storage, the areas stressed in the report, we notice a modicum of realism. The report's authors admitted the demand for enrichment wasn't likely to grow beyond existing capacity. It stated that demand was 23.5 million separate work units (SWU)[2] and that it was "expected to grow" to 29 million SWUs by 2025. However, enrichment capacity was already 39 SWUs. Its case for "potential" involvement in enrichment was not based on growing demand but on the facilities of the U.S. Department of Energy becoming "technologically obsolete" and on the fact that the diffusion process used was "tremendously energy intensive." By implication, this was an admission that nuclear energy contributes to greenhouse gases. (When nuclear is self-interestedly promoted as a magic bullet for global warming, I point out that all of Saskatchewan's uranium exported to the U.S. is enriched by two "dirty" coal-fired plants in Paducah, Kentucky.)

The report acknowledged that both France and the U.S. were involved in developing new enrichment methods. In the case of the U.S., the English company Urenco was proposing to build an $800 million centrifuge plant in Louisiana but is now considering such a plant in New Mexico. Further, the report acknowledged that two producers (Urenco and the Russian company Tenex) were already utilizing the centrifuge process, which "requires only about 2% of the electrical power of a diffusion plant." Urenco was doing enrichment for the U.K., Holland and Germany, whereas Tenex was doing it for the former U.S.S.R. Areva of France and the U.S. Energy Corporation are the other companies involved in uranium enrichment.

Even so — and not asking why nuclear powers like the U.S. and France would ever consider depending on Canada to do their uranium enrichment — the AECL consultants continued to discuss Saskatchewan's "competitive potential." The basis for this "potential" was primarily that, "There is well organized opposition by environmental groups who have filed for intervener status.... [and] If the licensing process in Louisiana is ultimately not successful then Saskatchewan might be considered as an alternate site." Showing the wishful thinking that comes from such a blatantly promotional report, the consultants went on, saying, "Alternatively, in the longer-term Saskatchewan could become the site of a second North American centrifuge plant." However, when discussing the state of the enrichment technology being explored for Saskatchewan, the AECL-commissioned report admits it "is not a proven process."

Environmental opposition in the U.S. and an unproven enrichment technology in Canada were hardly reasons to seriously consider a facility in Saskatchewan, especially when you consider the *de facto* monopoly over enrichment given to the U.S. in the 1988 FTA. At every point when an equally strong or stronger case could be made that there is little or no potential for such nuclear facilities in Saskatchewan, the opposite was concluded.

The purpose of the exercise was to create a public image of economic boom, as had been attempted during the late 1970s. In retrospect very little of the promised benefits accrued from uranium mining, but the objective was to get the mines on-stream, and corporate exaggerations are cheap. So, even with shallow foundations for their analysis, the AECL consultants went on to speculate about how an unproven 250,000 SWU enrichment plant would cost $50–80 million, generating $25 million in revenue and employing "approximately 80 people" by 1996. (The hypothetical plant was to use the method called Chemical Reaction by Isotope Selective Laser Activation, or CRISLA.) In its far-fetched nuclear wish-world, these 80 jobs would cost over $600,000 each and perhaps a million dollars in total capital costs. At these costs, in a nuclear world, many more of us would end up unemployed.

Electrical Forecasting without Assessing Energy Options

The real corporate purpose of the AECL-commissioned report was to paint a picture of nuclear boom time, which made the construction of a Candu-3 in Saskatchewan appear inevitable. Thirteen of the thirty pages on the "potential" for the whole nuclear fuel system dealt with nuclear power. But the "analysis" of electrical generation and the argument for the construction of three Candu-3s in the province was fraudulent, based as it was on SaskPower figures prior to SaskPower's own public review panel reporting back. The recommendations of SaskPower's Energy Options Panel did not bode well for the AECL or its Candu-3.

In October 1991, before SaskPower's Energy Options Panel reported in November,. the Devine government hastily signed a Memorandum of Understanding (MOU) with the AECL to consider building a Candu-3 in the province. This irritated SaskPower's Panel because such a "pre-emptive strike" could compromise the panel's independence and credibility. That this is the AECL's way of "doing business" should be kept in mind in considering the AECL's position that Saskatchewan needed a series of Candu-3s.

Let's follow their nuclear ideo-logic step by step. The AECL consultants' report noted that coal and hydro produced 95% of the 2,900 megawatts (MW) of electrical generation in Saskatchewan. If the capacity of the upcoming Saskatchewan coal-fired Shand Power Plant (280 MW net) was added in, with the expected loss of capacity and defunct import contracts deducted, the authors concluded the total capacity in 1992 would have been 3,050 MW.

The report talked of the need for ten, and perhaps fifteen, years' lead time for designing, approving and constructing new generating capacity and how environmental reviews are slowing down the overall process. The factors they considered in "forecasting future electrical energy demand" included economic growth, demographics, economic development and "other" macro economic trends, but there was no mention whatsoever of the technological revolution occurring in conservation and energy efficiency.

The rationale for their projections was astonishing. After they took forecasts from SaskPower's dated report, *Our Future Generation*, they said,

> To assess the reasonableness of these projections, we have compared them to the results from a simple forecasting model developed by

> Peat Marwick Stevenson and Kellogg. Our model is based solely on the relationship between Provincial Gross Domestic Product (GDP) and electricity demand; this relationship has historically been one of the best predictors of electricity consumption.

The concurrence between the two "forecasts" for the year 2000 (around 3,400–3,450 MW) was taken as some sort of reliability and validity test. Yet there was widely available research that fundamentally questions the need for there to be such a direct relationship between economic growth (GDP) and electrical demand (MW). For example, David Brooks's *Zero Energy Growth for Canada* had been available since 1981. Increased energy efficiency and co-generation in the industrial and commercial sectors, which are responsible for the major increases in electrical demand, can decouple economic growth from energy growth. Such information, in the September 16, 1991, feature, "Conservation Power," in *Business Week*, was readily available to the AECL consultants.

In contrast, the Energy Options Panel stated that "load forecasts, by their nature, are subject to a considerable degree of uncertainty, since it is often difficult to appreciate and quantify all the factors that influence future electrical energy and power requirements."[3] It went on to say that "demand side management (DSM) strategies and initiatives may provide a means whereby some of the risks associated with forecasting and planning new capacity additions can be reduced." This point is also stressed later in the report: "The DSM programs enhance power system planning flexibility since they can adapt relatively quickly in response to uncertainties in the load forecast and customer energy use behaviour."

The AECL's energy consultants seemed oblivious to these considerations. On this basis alone, it was apparent that they were serving a partisan role as "industrial propagandists." Their foregone conclusion was that there would be a 400 MW "shortfall" in Saskatchewan by the year 2000, which fit well with the AECL's agenda of building one "prototype" 450 MW Candu-3 during the 1990s. The closing down of the Queen Elizabeth coal station in Saskatoon after the year 2000 would, they claimed, require the building of yet another Candu-3 by the year 2005. Their "deterministic" projections are exposed by the fact that, at the time of this writing in the spring of 2007, there is no Candu-3 reactor under construction nor in the long-term plans of SaskPower.

The AECL consultants went through a ritualistic comparison with other "energy options"; however, their purpose was to make a case for their AECL bosses that Saskatchewan truly needed the AECL's Candu-3. Their discussion of demand side management, importing power and other sources of new generating capacity lacked fundamental credibility. The leadoff sentence on page 20, "To meet the future electrical requirements of Saskatchewan residents and industries there are three options," sets the tone. The commercial/industrial sectors, not the residential, are primarily responsible for increases in electrical demand. The listing of "residents" before "industries" leaves the false impression that families and homeowners are vulnerable without nuclear energy; a soft peddling of the "you'll freeze in the dark" fear-mongering we saw in the 1970s.

The AECL consultants argued that SaskPower DSM programs (Warm-Up, Enerwise, Energy Audits, energy efficient devices, and street and farm light conversion) had already been taken into account in SaskPower forecasts. While they admit information campaigns, customer rebates and incentives to conserve were being considered, there is no accounting of further reductions in electrical demand resulting from these programs. Furthermore the AECL consultants failed to explore technological improvements in electrical end-use, as well as new government policies and programs addressing other means of reducing demand and increasing efficiency.

There was great irony and hypocrisy in all this. The report acknowledged that SaskPower says that, "a further reduction of up to 200 MW in Saskatchewan's demand growth might be feasible." But they rejected this option because it would require "enormous capital investment in new technologies." Such an "enormous investment" is, of course, exactly what the AECL was advocating with the Candu-3. And study after study the world over, for example, Amory Lovins' work, show that it is far more cost-effective to invest capital in technologies that reduce demand rather than those that increase supply. Further, as discussed previously, if full costing is done of such things as uranium tailings management, reactor decommissioning and nuclear waste storage, then capital costs for nuclear power are far higher than for other sources of energy. The AECL actually admits nuclear power is extremely capital-intensive, yet when the case is made that far less capital is required to develop technology to reduce demand, they become pro-nuclear "Luddites," opposing out of hand other energy technology innovation.

What was SaskPower's track record in reducing demand for electrical energy or what is called demand side management? According to the findings of SaskPower's Energy Options Panel in 1991, SaskPower's conservation programs saved "approximately 56 gigawatt hours per year and approximately 14 megawatts (MW) in demand." The cost of this was about $14 million, and some of these programs could have had a much greater impact if the Devine government had not cut them in the early 1980s. The exception was the government's expansion of natural gas into rural areas, which enabled farm and rural residents to convert from electric space heating. With natural gas supplies declining, this is only a short-term solution; nevertheless the Energy Options Panel reported SaskPower's estimate that this presented "potential savings of 330 MW compared to their collective use of electric heat." This alone would undercut the shortfall being projected by the AECL consultants as a reason for building a Candu-3. In the previous decade (1980–90), according to the Panel, SaskPower estimated that DSM saved "about 400 MW of demand and 600 gigawatt hours (GWh) of electricity." The Panel reported that from 1990 to 2000 SaskPower said "current and proposed DSM programs are expected to reduce growth in demand by 150 to 200 MW from growth that would have occurred in absence of these programs." Over twenty years a saving of 22,000 gigawatt hours, "equivalent to two years of current energy sales to the entire province," was expected.[4]

In sharp contrast, the AECL downplayed further potential for DSM. Their consultants' report said that trying to get "a further reduction of up to 200 MW in Saskatchewan's demand growth might be feasible but would require either substantial investment by consumers or significant changes in lifestyle." The mention of "lifestyle" is a subtle appeal to the consumer society's addiction to mindless "convenience." Energy conservation and renewables would actually reduce demand and enhance "quality of life." However, such a reduction of demand for 200 MW would be fatal for the AECL's plans, as it would reduce more than one-half of the projected 400 MW growth that the AECL wanted us to believe necessitated the construction of their Candu-3 by the year 2000. It was hypocritical and nonsensical for them to stress the "substantial investment" required for DSM, when investment in a Candu-3 would be several times greater.

SaskPower's past efforts and future plans for DSM didn't compare well to experience elsewhere. The Energy Options Panel pointed out that one U.S. utility plans "at least 13,000 gigawatt hours of conservation and energy efficiency improvements by the year 2000. This represents over half (50%) of the projected energy growth required in a medium/high growth scenario for the region." It pointed out the costs of this would be "lower than new generation." The Panel's final report noted that this compares to SaskPower, which "anticipates being able to meet about seven percent of projected energy growth during the next decade from DSM" (i.e., reducing energy growth from 3% per year to 2.8%).

Another example of DSM was reported in the September 16, 1991, *Business Week*. California Edison needed 6,400 megawatts by 2010, which is about thirteen times the expansion that the AECL consultants said Saskatchewan needed by 2000, but Edison wasn't going to build a nuclear plant. Instead the new capacity would come from upgraded plants, and some from independent producers of solar and geothermal energy. And 4,400 MW, or 70% of the total, would come from conservation, from "freeing up electricity that now is used inefficiently."

Energy Options noted that world-renowned energy efficiency researcher Amory Lovins of the Rocky Mountain Institute estimated twice as much electrical energy could be saved as just five years before, in 1986, and at one-third the cost. And the best energy-saving technologies weren't even available a year before. It's little wonder that the Energy Options Panel's first recommendation was to study and improve energy conservation and efficiency in SaskPower. It wrote:

> The potential for economically feasible conservation and efficiency improvement in Saskatchewan should be carefully and exhaustively evaluated. In the absence of such a study, it is difficult to predict the extent to which conservation and efficiency improvements will impact upon future electric power and energy requirements. SaskPower should conduct a thorough review of demand side programs in other jurisdictions and their relevance to Saskatchewan.

Meanwhile SaskPower officials were downplaying the potential for DSM. Could this have been related to them signing an agreement with the AECL to develop a Candu-3 even before their own Energy Options Panel reported to the Crown corporation? It is well known that Romanow-appointed SaskPower president Jack Messer, who was Minister of Mineral Resources when uranium mining expanded in the late 1970s, was a staunch supporter of the nuclear industry, and in this regard he wasn't very different from past SaskPower president, Tory-appointed George Hill. After the 1991 NDP provincial convention passed a resolution by an 80% majority calling for cancellation of the SaskPower-AECL Candu-3 agreement, while talking in the rotunda of the Regina Centre of the Arts, Messer said directly to me that he didn't have to do what the party wanted. Non-nuclear activists speculated that Messer would try to get the AECL-SaskPower agreement approved by the March 30, 1992, deadline, prior to updated knowledge of the potential for DSM becoming widely available, but as we shall see later there were some surprises in store for the nuclear proponents.

The AECL, its consultants and its allies in SaskPower clearly didn't want Saskatchewan people to know the incredible potential for energy conservation and efficiency. In combination with renewable energy and co-generation, systematic DSM would quickly rule out the prospects for a Candu-3 in Saskatchewan. However, since realities and facts hadn't stopped the nuclear industry in the past, the people of Saskatchewan had to remain on guard.

Deconstructing AECL's "Public Acceptance" Strategy

There's a much more insidious side to this integrated uranium/nuclear agenda: the AECL and Cameco's attempts to make Saskatchewan the international site for disposing highly radioactive nuclear reactor wastes. As I have already discussed, the AECL's public credibility remains low due to the buildup of reactor wastes at Candu reactors, primarily in Ontario. The nuclear industry approaches reactor waste "disposal" with the same bravado, technological naiveté and lack of foresight with which it embarked on the construction of nuclear weapons and nuclear reactors in the first place. It continues to hide behind fail-safe or low-risk models, while downplaying the technical and ecological downsides. Though the 1991–98 FEARO Panel looking at the AECL's deep rock nuclear waste disposal didn't encourage this, the industry-based NWMO has returned to these dangerous myopic practices. The AECL prefers to focus on the "public acceptance" side of their nuclear waste problem. About this, the AECL-consultants wrote, "If the Saskatchewan public is prepared to accept a Candu-3 in the Province's electrical generation mix, this means, at a minimum, that used fuel would be stored at the reactor site. It could *then be argued* that developing a waste repository to ultimately dispose of used fuel in Saskatchewan *is the responsible course of action*" (emphasis added).

It is revealing how, in the AECL's ideo-logic, one Candu-3 becomes the basis for Saskatchewan taking responsibility for massive reactor wastes from elsewhere. This shows how the Candu-3 proposal was a nuclear Trojan Horse. Strangely enough, the fact that over 90% of Candu reactor wastes in Canada come from Ontario Hydro's plants hasn't yet made Ontario responsible for permanent storage of these wastes.

But this was only the beginning of the AECL's plan. Its consultants also wrote: "A used fuel disposal centre dedicated to two Candu-3 reactors would be prohibitively expensive in comparison to a national repository.... [Therefore] for Saskatchewan to become a national high level waste repository the public must be willing to accept used fuel from Candu reactors in Ontario, Quebec and New Brunswick being shipped to northern Saskatchewan for disposal." Orwell would be impressed by the turn of phrase.

The AECL's "public acceptance" strategy continued to unfold in the report: "This could initially be tied to used fuel derived from Saskatchewan uranium.... Once a waste disposal program is firmly in place it could draw from a broader market including Candu-3s exported to third countries, customers of Saskatchewan uranium, and light water reactors generally. Fuel leasing arrangements could be considered where uranium would be provided and the used fuel taken back for ultimate disposal."

The AECL-commissioned proposal was to transport highly radioactive fission and transuranic products (like plutonium) across Canada by rail and road. This would be done, the authors said, in 35 tonne casks engineered by Ontario Hydro, each containing 192 spent fuel bundles. If each bundle weighs approximately fifty pounds, each cask could take around 9,600 pounds of spent fuel. In 1990, Candus in Canada created about 1,800 tonnes (or about 3.6 million pounds) of radioactive spent fuel, so just to keep up with this steady supply would require 375 shipments a year, or a little more than one per day for as long as these reactors were operating.

Without any nuclear waste "disposal" system in existence, radioactive spent fuel has steadily been building up at Candu reactor sites. The FEARO Panel that started hearings on nuclear waste disposal in 1991 estimated there were 40,000 tonnes (or 80 million pounds) of such wastes in Canada. Using the AECL consultants' plan, this would require 8,333 shipments to a nuclear waste repository, which would be twenty-three per day for a year, or 2.3 a day for ten years. However, even accepting the much lower (and erroneous) AECL consultants' estimates that by 1989 there were at least 14,000 tonnes (or 28 million pounds) of spent fuel stored at Candus in Canada, there would still be 2,916 shipments of radioactive spent fuel, or eight per day for a year, or one a day for eight years.

If the AECL were to get its way and build a Candu-3 in Saskatchewan as a step towards creating a nuclear waste dump here, we could be faced with daily shipments of the most radioactive substances on earth for decades to come. These shipments would be going through our towns and cities 365 days a year, year-in and year-out for long after these reactors were decommissioned. Imagine the magnitude of radioactive wastes going through Saskatchewan if our north became a continental and/or international nuclear waste dump. It is unthinkable.

Despite glib promotional guarantees by industry-employed "experts," nothing in the nuclear industry is fail-safe. Saskatchewan would face the threat of frightening rail or truck accidents. If other places don't want these nuclear wastes and these risks, is there any reason why we should want to take them? Again, notice the self-serving ideo-logic. The nuclear consultants have gone from constructing a Candu-3 — unnecessary in order to meet Saskatchewan's electrical demands — to us being responsible for permanent disposal of the reactor's waste, a responsibility not accepted anywhere else. The consultants next recommended accepting all of the nuclear waste from across Canada to create an economy of scale, even though the waste is created elsewhere, mostly in Ontario. They then suggested agreeing to accept all nuclear wastes involving uranium fuel from Saskatchewan, and ultimately nuclear waste not involving uranium fuel from Saskatchewan. This might include accepting nuclear reactor waste from light water reactors in the U.S., where the industry is facing stiff opposition to locating a nuclear waste site at Yucca Mountain, Nevada. It might include taking the nuclear wastes from France's huge nuclear power system. What a deal for Ontario Hydro, for the U.S., for France and for the worldwide nuclear industry! And all happening because of the generous spirit of Saskatchewan people, whom the nuclear industry clearly see as desperately naive and gullible.

Even the AECL's arguments for taking nuclear reactor wastes from across

Canada and abroad contradict official nuclear ideology. The position of the AECL has consistently been that the export of its Candu, and the provision of uranium from Canada, has not increased the risk of nuclear weapons proliferation. The Non-Proliferation Treaty and international inspection, it has argued, protect us all from this threat. But all of a sudden Saskatchewan should take back nuclear reactor wastes from all customers of the uranium industry in order to "reduce the risk of nuclear proliferation." It seems no energy, environmental or nuclear proliferation crisis is beyond the exploitative gall of the AECL. But that's not all. In the AECL's dark future, Saskatchewan might also become the site for other toxins, for the AECL consultants wrote, "It may also be practical and commercially lucrative to use the repository for disposal of other highly toxic wastes which cannot be destroyed with conventional waste management practices."

The AECL consultants listed several "factors which favor Saskatchewan" as a nuclear waste site. They mention the Precambrian shield and that the north "is sparely populated and that population is already familiar with the nuclear industry because of uranium mining." The argument had come full circle. The AECL liked the fact that an infrastructure has been developed in northern Saskatchewan, due to uranium mining, which could also be used to justify disposing nuclear wastes.

When uranium mining was expanded in the north in the late-1970s, the stage was being set for a national and global battle over nuclear reactor wastes. In retrospect, uranium mining was itself a nuclear Trojan Horse. We are confident that Saskatchewan people committed to ecologically sustainable development, non-proliferation and the well being of future generations will have something "big" to say about these less than forthcoming and dangerous plans of the nuclear industry.

Notes

1. R. Lloyd, J. Bonny, and J. MacDonald, *Prospects for Saskatchewan's Nuclear Industry and Its Potential Impact on the Provincial Economy, 1991–2020*, Saskatoon, SK: Peat Marwick Stevenson and Kellogg, Oct. 21, 1991.
2. SWU is a unit of measurement of the enriching effort needed to separate the U-235 and U-238 atoms in natural uranium to create a product richer in U-235 (about 4–5%) that can be used for fuel in nuclear power plants.
3. SaskPower, *Energy Options Panel, Final Report*, Regina, Saskatchewan, November 1991.
4. SaskPower doesn't have the actual figures, but all indications are they didn't come close to maximizing DSM during this period. Peter Prebble, who prepared a 2006 report for the Calvert NDP government on energy conservation, says that SaskPower spends less than $1 million a year on DSM, compared to twenty-five times this amount spent in Manitoba and B.C., where DSM is treated as a capital expense and competes with other supply options (personal communication, Nov. 14, 2006).

chapter 13

NUCLEAR TROJAN HORSE

Bribing Us with Our Own Money

Nuclear power became a contentious issue in the 1991 Saskatchewan provincial election. The Devine-led Tories had just signed a Memorandum of Understanding (MOU) with the AECL, and the Linda Haverstock-led Liberal Party also strongly endorsed a Candu nuclear reactor industry as an economic development strategy. (Later Haverstock became the second pro-nuclear person in a row to be appointed as Saskatchewan's lieutenant governor.) Somewhat belatedly and sheepishly, feeling the need to distinguish itself from the other parties and appeal to its base of support, the Romanow-led NDP mentioned the potential of energy conservation without actually getting into energy policy in any depth. As though trying to avoid the issue, the NDP made no mention of its long-standing policy supporting the phase-out of uranium mining nor of the connections between the Candu-3 and the military-industrial complex.

The municipal and provincial elections overlapped in 1991, and, while running for Regina City Council, I tied to raise the local profile of the nuclear reactor debate. With an explicit anti-nuclear position, I came twelfth of forty-eight candidates in 1991, but only the top ten were elected to council. Running again in 1994, on a left green program stressing ecological preservation, Aboriginal justice and inner city revitalization, I was elected to represent Regina's inner city ward.

By 1991, nuclear power was already a vital issue in Saskatoon, which was the Canadian headquarters for several multinational uranium corporations. In this corporate-influenced climate, in 1987–88, the Inter-Church Uranium Committee had waged and won a heroic campaign to have Saskatoon designated as a "Nuclear Weapons Free Zone." However, civil and business leaders narrowly interpreted the declaration to exclude the weapons connections with the uranium industry.[1] Though more removed from this corporate fray, nuclear power was becoming an issue in Regina too. Half-truths and outright distortions about nuclear power being economically prosperous and environmentally benign were covertly promoted throughout our city without independent and objective views receiving any attention. We were being propagandized with our own money.

Realistically Assessing the Candu's Costs and Benefits

With provincial politicians and city and town business lobbies toying with nuclear power, there was an urgent need to take a realistic look at the exaggerated benefits and unmentioned burdens of the industry. Under the Devine Tories, the province had accumulated more than $5 billion of debt in less than a decade, and SaskPower was carrying a debt of over $2 billion. If Saskatchewan "went nuclear," it would be a further drain on the provincial budget, shifting scarce resources from the municipalities and essential services to capital and interest costs and high-paying technical jobs.

Cost overruns in the nuclear industry already placed a serious burden on taxpayers. The AECL estimated the cost of building its new Candu-3 at $1 billion, which was an increase from just a few years before. Based on Ontario's experience, the cost overruns could be two or three times that amount. Furthermore, as Ontario's nuclear reactors age and face more serious safety problems, the hidden costs of the industry become exposed, and the myth of nuclear as a cheap source of energy is laid to rest. On September 21, 1991, as reported in the Regina *Leader Post*, Energy Probe researchers pointed out that Ontario's nuclear utilities cost taxpayers $1 billion a year, and Ontario Hydro had recently raised its rates another 12% and expected to make increases annually into the 1990s to try to reduce its debt load and cover repair costs.

There was a plethora of other issues not mentioned by the nuclear proponents. As we've already seen, nuclear energy isn't a realistic or cost-effective strategy for reducing greenhouse gases and global warming, and furthermore the nuclear industry uses massive fossil fuels in uranium mining, uranium enrichment and the other energy-intensive parts of the nuclear fuel system — fossil fuels not needed with renewable energy. Also, nuclear energy does nothing to address the role of transportation and deforestation in global warming, and while focusing on pollutants from other energy sources, nuclear proponents ignore their own waste problem. If long-term waste management costs were factored into the price of uranium and the cost of nuclear power, the industry would instantly be driven out of the market and wouldn't be able to play on the hopes and fears of Saskatchewan people.

There should be no illusions about the serious implications that a nuclear industry would have for Saskatchewan. By the 1990s, we had already seen a massive shift of government resources from the agricultural economy to the northern mining economy as part of the expansion of uranium mining, and nuclear power would continue the de-development of rural Saskatchewan. Meanwhile, very few direct short-term benefits went to northern Indigenous people — only about one-sixth of the jobs they were led to believe would come. Nurturing renewable energy, like wind farming across the Prairies, would create far more direct employment and diversify agriculture in an ecologically sustainable way.

But this industry has no lack of trickery up its sleeves. Exaggeration, especially when many people are down and out, as in the north, has always been one of its ploys. So, before civic leaders became bamboozled by a gold-rush mentality, they needed to be challenged to take a hard look at the earlier "uranium boom" and the benefits actually received by the province — a pittance compared to what had been promised. By 1989 the province was to have received as much as $467 million from the uranium industry. Even the lowest estimate for provincial revenues by then was $196 million. In actuality, the province got only $15 million.[2] John Warnock's research has shown that uranium revenues going to government as a share of corporate revenue have continued to fall since then.[3]

The government spent much more than it received to develop the uranium industry. Cameco was created from two Crown corporations, the SMDC and Eldorado, with massive support from the taxpayer. It was privatized under the federal and provincial Tory governments in 1988, at a time when its market value was declining due to the glut in the world uranium market and bargain-basement uranium prices. Though some individual shareholders benefitted financially, the public will never get back the many hundreds of millions of dollars that Saskatchewan governments invested in this corporation over the years. In 1994–96, and later in 2002, when the Saskatchewan Government sold 9.5 and then 5 million Cameco shares, it did get some investment back.[4] In February 24, 1992, after a lengthy investigation, Paul Hanley, the environmental reporter for the Saskatoon *Star Phoenix* concluded that, "the investment by taxpayers was three times as large as the taxes and royalty revenues received, and this excluded many of the indirect costs."

Ironically, Grant Devine wasn't far off the mark when he campaigned against uranium mining in the 1982 provincial election. But the memories of politicians are short. Still in opposition, he was critical of the squandering of provincial money, an estimated $500 million by 1982, because, as his campaign literature stated, "There is no return on that investment yet and the price of uranium has fallen by more than a half." The price of uranium went down even further in his term of office.

The NDP's original plan was for some of the projected massive uranium revenues to go into the Heritage Fund, which would help finance essential services, including grants for municipalities, and cover some of the costs of decommissioning uranium mines. We've heard nothing about this fund for more than a decade because the Heritage Fund was depleted by vote-getting programs and the subsidization of the uranium industry.[5]

We heard more boom-town exaggerations in the 1991 provincial election. In his first TV press conference Devine bragged that the nuclear industry would bring 15,000 jobs to Saskatchewan. Later, as reported in the September

14, 1991, Regina *Leader Post*, SaskPower president George Hill upstaged this by saying that 30,000 to 35,000 jobs would come with the nuclear industry. When other less-politicized SaskPower officials discussed their joint venture with the AECL, the figure was lowered to 6,000 jobs. This, too, was ridiculous, as the figure was similar to the number of short-term jobs created by the construction of the Darlington nuclear plant in Ontario, which was at least eight times the scale of the proposed Candu-3 reactor in Saskatchewan.

In actuality only 170 jobs were guaranteed by the SaskPower-AECL agreement, at a cost of $50 million to the taxpayer. Nothing more was guaranteed. And these jobs mostly involved transferring highly paid technical and bureaucratic AECL jobs from elsewhere, rather than a net gain in jobs. This was akin to deals cut between the provincial and federal Tories over the federal agricultural program Fair Share, and the financial company Crown Life, which also shuffled shrinking jobs from the east to the west.

It's Hard to Carry a Trojan Horse on a Nuclear Tightrope

SaskPower walked a nuclear tightrope with its AECL deal. Its officials claimed there was no commitment to nuclear energy until its own Energy Options Panel reported. Meanwhile, the AECL's Candu president was reported in the September 21, 1991, Regina *Leader Post* as saying Saskatchewan had an "obvious need" for a reactor. The AECL's new Saskatchewan-based vice president, David Bock, was more honest when he said in that same day's *Globe and Mail* that the AECL needs "a place to demonstrate it." All the while both SaskPower's and the AECL's projections for Saskatchewan electrical growth were being challenged by the National Energy Board (NEB). While the former said our growth would be above the national average, the NEB said it would be below the national average. The NEB's projections were later confirmed by an internal NDP government study, reported in the March 4, 1992, Regina *Leader Post*. Self-interest was running rampant in the AECL's search for a place to locate its Candu-3 Trojan Horse.

SaskPower president George Hill had already gone on record in the September 14, 1991, Regina *Leader Post* as embracing the AECL's scenario that, with a homegrown nuclear industry, Saskatchewan should also become a dumping ground for nuclear reactor wastes from abroad. It was the AECL's board chair Robert Ferchat who first suggested the ludicrous idea that it was somehow in the interest of Saskatchewan people to "lease" uranium and then to take back the deadly reactor spent fuel, including plutonium, as part of the deal.

Even if it had proven true (which it wasn't) that 400 megawatts more generating capacity was needed in Saskatchewan by 2000 and that nuclear power was the best option, surely a joint venture with the AECL should have waited until the Energy Options Panel reported back. Otherwise why did

SaskPower go to all the trouble of having this expensive public review? It is likely no coincidence that the MOU was signed just hours before Devine called the provincial election, suggesting the whole thing was a bribe aimed at civic and business leaders. Had the Tories won, we might have had to directly oppose a Candu industry starting in Saskatchewan. Electing the NDP — which acts more like the Nuclear Development Party than the New Democratic Party — probably helped stall the nuclear expansionist strategy in the short term.

The AECL and the Devine government clearly used SaskPower for their own commercial and political purposes. According to the September 21, 1991, *Globe and Mail*, the MOU with the AECL said that if SaskPower decided not to proceed with the Candu-3 project, the AECL was obliged to repay the $25 million SaskPower had contributed — a no-risk inducement for being a public partner, or, perhaps more accurately, a "co-conspirator" with the AECL.

Struggling for credibility, SaskPower officials tried to make this MOU look like it wasn't a back-room deal. They stressed that site-specific studies and environmental reviews still had to be done. But the public already knew from what happened with the Rafferty-Alameda dam project that if a federal review panel had serious problems with a Saskatchewan project, the Devine government was willing to proceed anyway. And, under NDP governments, the Saskatchewan environmental review process hasn't had much chance of stopping any mega-project that was supported by the political warlords. From 1980–90, Saskatchewan had over six hundred proposals referred for environmental assessment, and only 13% of these were judged to need an assessment. Only two of these were referred to a review.[6]

The nuclear industry and its political backers inflate the short-term pay-offs of nuclear power, ignore the hidden costs to the taxpayer and push massive ecological costs to future generations. As already noted, the January 16, 1989, *Globe and Mail* reported that the Economic Council of Canada estimated that $12 billion in government subsidies had already gone into the Canadian nuclear industry. The figure now is closer to $20 billion without considering inflation and interest on debt. With this kind of government backing it's not surprising that the AECL could afford to give away thousands of expensive, glossy promotional pamphlets in malls and schools in Regina, Saskatoon and across the province. This was a sober political lesson for the Saskatchewan electorate and taxpayer, as this manipulative, publicly funded wastefulness occurred in the midst of growing child and family poverty in the province. Ironically, most of the increases in poverty were among Indigenous families, the very people that the "uranium boom" was supposed to help "get ahead."

And there are more insidious forms of public subsidy. At the time,

Ontario's Energy Probe was preparing a case against the nuclear industry for its lack of insurance for corporate liability. Under the Canadian *Nuclear Liability Act* the industry is not liable for deaths, injuries and ecological damages when a reactor accident occurs. It is worth remembering that, despite all the public pronouncements about Candu safety, in its October 16, 1989, submission to the Treasury Board, the AECB regulators said, "Candu plants cannot be said to be either more or less safe than other types." Under this Act, utilities like Ontario Hydro and SaskPower (had it gone ahead with a Candu-3) have a cap on their liability to protect them from bankruptcy. No other industry has such absolute protection from liability. Even the chemical industry, with its litany of toxic products, is ultimately liable.

Civic and local business leaders lured by nuclear development might ask themselves, if the nuclear industry is as safe and clean as it claims, why is it not willing to purchase its own private liability insurance? With the cost of accidents and private insurance not borne by the industry, the temptation to loosen safety standards or take short cuts is impossible to ignore.

If the SaskPower-AECL agreement had led to the construction of a Candu reactor in Saskatchewan, its financial, economic and environmental effects would have quickly filtered down to contribute to a further decline in the quality of life in our towns and cities. Public services and grants to municipalities would have been cut even more than they were by federal and provincial off-loading. The province's desperate fiscal situation when the Romanow NDP took power in 1991 may even have been its bottom-line reason for not proceeding with some version of the AECL Trojan Horse.

We might ask why so many municipal politicians at the time remained silent on a matter that had such dire consequences for their cities and towns. Rather than believing in the golden egg, we need to keep our eyes firmly fixed on the financial big picture, so that temptations won't be blinding if NDP Premier Calvert or a future Saskatchewan Party government again tries to build a uranium refinery or nuclear power plant in the province.

NDP Victory Convention: Not the Expected Love-in

The Saskatchewan NDP had a provincial convention scheduled in the wake of its October 21, 1991, election victory. Most observers expected an uneventful love-in. Many party activists were willing to postpone contentious policy debate until the new caucus and Cabinet got their teeth into governing. The uranium and nuclear industry lobbies, however, did not want to lose the momentum they had achieved under the Devine Tories. They still wanted to get new uranium mines on-stream and a Candu-3 industry up and running in the province.

SaskPower's Energy Options Panel had issued its final report in November 1991. It placed conservation and accurate forecasting as a prerequisite to any consideration of new power plants, which did not bode well for the time-line and strategy of the nuclear proponents. The proponents, however, had been hard at work with their inner NDP allies and caught other party delegates off guard. The ensuing clash provides stark insight into the manipulative nature of party politics within a governing party and shows how inconsequential long-standing NDP policy became in the Romanow period of rule.

I attended the convention panel that debated environmental issues and took extensive notes. There was unexpected high drama as NDP delegates convened only three weeks after the party's return to government. The goal of party unity going into the first year of a return to government was espoused by the party's well-organized Environmental Caucus, which had strongly endorsed the party's policy against nuclear power and for the phase-out of uranium mining. But caucus members soon became aware a resolution had been submitted to reverse the party's eight-year opposition to uranium mining. The Environmental Caucus offered a "deal" to the NDP president and the pro-nuke lobby to table all the resolutions if there was a fair, year-long, grassroots educational process on these matters and an orderly debate at the 1992 convention. (The party had undertaken such a process in 1983, which had led the party to reverse its position from a pro-uranium to a phase-out stance.) The deal was rejected outright, "from the top," purportedly because of a concern that such a high profile process might draw attention from the media. Environmentalists in the party speculated that the pro-nukes knew they would lose ground in an open educational forum, that party democracy was not on their side.

With only minutes left before the panel "The Economy and the Environment" was to begin, the fight was on. (Note how these topics — Economy and Environment — are combined, which can favour the ideology of capitalist economic growth and jobs at any cost.) A group of delegates from the Steelworkers union and their supporters came into the foyer at the Regina Centre of the Arts like a football team storming a high school party. They were there to ensure a go-ahead on a new rash of uranium mines coming up for public review, and you could sense they meant business.

In two short hours the panel had to debate twenty-six resolutions that spanned environment, labour and social issues, nuclear and uranium, and economic policy. The tactic of these Steelworkers was to try to move the resolution favouring the expansion of uranium mining to the front of the debate to ensure it would get addressed.[7] Two-thirds support was required to have this pro-uranium resolution move forward. On the first vote, by a show of hands, support looked strong. By the time a standing vote was called for and tallied, and people who had been milling around unaware of what was happening finally got into the panel, the vote was lost.

The Environment Caucus may not have fully mobilized its province-wide support for non-nuclear resolutions at this convention. However, it was becoming clear the pro-uranium forces, primarily within that part of labour with short-term benefits from high-paying mining jobs, came prepared to get their way. Because many environmental delegates — in other panels running concurrently, such as the one on Aboriginal self-government — were slow to enter this panel, the nuclear proponents within the party came close to winning.

There was already some indication, later to be confirmed, that the inner cabinet supported the tactic of the Steelworkers. Ed Tchorzewski, newly appointed deputy leader and Minister of Finance, was openly supportive of the motion to bring forward the pro-uranium resolution. The first resolution on nuclear/uranium issues to come up was a general one making conservation the priority over mega-projects like nuclear or coal. Terry Stevens, head of the Steelworkers, quickly moved an amendment to remove the last part of this resolution, which stressed that this should be done "before any further public debate over nuclear or coal fired generation of electrical energy." He stated his motive for doing this was that he didn't "want to be gagged" about further public debates about nuclear power. His implicit appeal to freedom of speech was effective, and the amendment carried.

Another delegate moved a resolution that improved the wording to: "before any further *commitment* to the development of nuclear or coal fired generation of electrical energy." This carried easily, showing the importance of such distinctions in winning support. It was noteworthy that Terry Stevens did not support this resolution, suggesting the appeal to "free speech" was a ploy and that the underlying concern wasn't about energy policy but prospective jobs in the nuclear industry. This near defeat, of even a general resolution, shows the need for careful wording of resolutions, so as not to give opponents the chance to challenge them on a seemingly semantic and emotional rather than substantive basis. Because time was running out, panel chair Bill Hyde ruled that equal time be given to the resolutions remaining on labour and social issues, and nuclear/uranium issues. This allowed fifteen minutes each, with those dealing with nuclear/uranium issues left to the very end, from 4:15 to 4:30 pm. During the debate on the labour and social issues, the tension continued to build, and the panel became jam-packed with delegates and observers because of the expected fight over nuclear and uranium policy.

The earlier debate on conservation proved to be a trial run for the first resolution to come forward on nuclear/uranium policy. This resolution directly opposed the 1991 Devine-Mulroney government's MOU, which referred to the possibility of a Candu-3, Slowpoke reactor and nuclear waste disposal site in Saskatchewan. Once this emotionally charged matter was brought into the discussion, it became clear that even without mobilizing there was momentum for the non-nuclear delegates. Long-time non-nuclear Saskatoon MLA Peter Prebble moved an amendment to update this resolution, calling for the cancellation of the AECL-SaskPower $50 million agreement signed just before Devine had called the provincial election. Another delegate was prepared to move an amendment that called for the $25 million share from SaskPower to go into energy conservation and related job creation, but because of time constraints this wasn't pursued.

Steelworker Terry Stevens again intervened, this time appealing to delegates to "trust Roy," the Premier, to do the right thing and not to tie his hands. Stevens clearly wanted the SaskPower-AECL agreement to stand. Someone from the panel heckled, "Leave it to Tommy," in reference to how often people (including Grant Devine himself) appeal to the post-mortem image and authority of Tommy Douglas, rather than the issue at hand, to try to get support for their views. The authoritarian deferential appeal to leader "Roy" failed, and Prebble's amendment carried with ease. With the help of well-reasoned support by Regina MLA Bob Lyons, past SaskPower critic, the overall resolution carried easily, by about 75%. The momentum had clearly shifted and time had run out on the Steelworkers and inner-cabinet supporters who wanted to reverse the party's uranium policy.

What might have happened if uranium policy was debated? It's possible that those not well versed in the overall issue but opposed to nuclear power might have supported reversing the party's uranium phase-out policy. But those who support uranium mining — like Terry Stevens, the Steelworkers and their inner-cabinet supporters — knew full well that these issues are interconnected. The only way to justify new uranium mines is with growing demand for nuclear fuel, and the AECL didn't have a good record of reactor sales over the last decade. Its strategy in targeting Saskatchewan over the previous two years was to get the province into a Candu-3 industry, in hope that this smaller reactor prototype would increase export potential. No matter what kind of lip service the pro-nukes pay to conservation, efficiency and renewables, they know that successes in these areas will directly undercut the demand for electricity from either coal and/or nuclear power. Like the Swedish social democrats, their strategy is to pre-empt rational planning, balanced incentives and the investigation of ecologically sustainable energy options.

The fact that conservation and a balanced evaluation of the energy options was evident in the Energy Options Panel report[8] released just before the NDP convention likely made party pro-nukes somewhat nervous and explains their push for a reversal of policy. They were willing to do this even though it would risk party unity and blemish the post-victory love-in. The pro-uranium resolution that they proposed emphasized business-as-usual — that rich uranium ore exists and that uranium is a fuel for electrical generation. It erroneously alleged that this source of electricity is the cost-effective option for developing countries and that it is an environmentally benign alternative to coal. This is identical to the industry position, which probably shows how effective the multi-million-dollar AECL and CNA campaign had been in certain sections of the party and labour establishment.

As it turned out, it didn't matter that the pro-nuclear lobby didn't get to debate or reverse the party's uranium policy — they were going to get their way regardless. In the aftermath of the post-election victory convention, new NDP Energy Minister John Penner indicated that the government wouldn't take guidance from the party's "phase-out" uranium policy, supposedly because it wasn't debated.[9] In other words, the inner cabinet rejected a year-long democratic process of resolving party conflict over uranium mining; the nuclear lobby in the NDP, trying to salvage the Tory-AECL Candu-3 agreement, attempted to steamroll through a reversal of the uranium phase-out policy; and after all else fails, the new NDP minister in charge declared that long-standing party policy — from 1983 to 1991 — was not to be followed because no debate on uranium occurred. Could there be a better example of why so many have become so cynical about party politics?

It's perhaps no accident that the pro-nuclear resolution came from the Kelsey-Tisdale riding, where the NDP's provincial election campaign manager and just-appointed acting SaskPower president Jack Messer resided. Messer, Minister of Mineral Resources under the Blakeney government, had a reputation for ignoring due process, having continued to promote uranium export contracts even while the CLBI was deliberating on whether the uranium industry should even expand. In light of his views and methods it is disconcerting that soon after the 1991 convention he became SaskPower's president. There's no way he would fairly implement policies of conservation or efficiency along the lines stressed by both SaskPower's own Energy Options Panel and NDP policy — demonstrating that patronage appointments by the Romanow government were no more likely to serve the public interest than had the patronage appointments of the Devine government. This was noted by journalist Dale Eisler in his November 16, 1991, Regina *Leader Post* column, "Real politics lack some nobility."

Candu-3 and the Military-Industrial Complex

In June 1991, the U.S. military multinational TRW[10] signed a secret Memorandum of Understanding with the Devine Tory government. In February 1992, the newly elected Romanow NDP government, apparently still looking for a way to salvage the AECL-SaskPower Candu-3 deal it inherited, sent a minister to visit the TRW in Washington, DC. Deputy NDP leader Dwaine Lingenfelter, who went on behalf of the government, later joined the oil industry and more recently has been lobbying for a Candu reactor to be built in the Alberta tar sands.

Briefing notes for that visit, which were leaked to me in March 1992, indicate that the AECL-SaskPower MOU was interconnected with the one with the TRW. Under it, Saskatchewan was earmarked to become partners with a military contractor in an interlocking corporate strategy involving the Candu-3 and nuclear waste disposal.

What follows is based on an article that appeared in the May 1992 news magazine *Briarpatch*, which was published anonymously. Due to the "climate of apprehension" at the University of Regina, where I was employed at the time, I did not feel "free" or safe to publish under my name.

The AECL's associate, the TRW, is one of the largest corporations in the U.S. military-industrial complex. The July 22, 1991, issue of *Defense News* ranked the TRW as the fifteenth largest military firm. Its sales that year were $8.2 billion, nearly half of which was from direct military contracts. The TRW is also involved in space research and hazardous wastes and, at the time, employed 70,000 staff world-wide. In the early 1990s it received a ten-year, $1 billion contract with the U.S. government for preparatory work for a nuclear waste dump at Yucca Mountain, Nevada.

And article in the March 1992 issue of *Harper's Magazine* reveals how the TRW worked alongside the American Nuclear Energy Council, an industry association like the CNA, and nuclear reactor utilities to overcome opposition by the Nevada public, which rejected this nuclear dumping site by a two-to-one margin.[11] Like the CNA in Canada, these nuclear groups were involved in a multi-million-dollar media campaign to influence the public about nuclear issues. In 1988 the U.S. Council for Energy Awareness spent over $8 million, which, according to its 1988 Recommended Media Plan, was "to maximize favorable opinions on nuclear energy among the target audience."

The TRW is at the centre of the military-industrial complex and is best known for producing Minuteman nuclear missiles. It has made huge profits from creating sophisticated weapons of mass destruction, which continue to threaten humanity with nuclear war while deflecting vital resources from the development needs of people around the globe.

The Science and Technology Division of the Saskatchewan Department of Economic Diversification and Trade began discussions with the TRW in August 1990. According to the January 7, 1992, "Ministerial Briefing Notes: TRW Visit," this department is credited with playing the "pivotal role in developing the AECL/TRW relationship...." Follow-up discussions indicated "TRW's interests in providing technology and systems to AECL as part of the proposed Candu 3 project... [and in] participating in the disposal of hazardous wastes." According to the addendum to the briefing notes, discussions were also initiated "via the Department between AECL and TRW to identify specific areas of co-operation."

The general manager of the TRW's Command Support Division signed the MOU with the Devine Tory government on June 5, 1991. The leaked Briefing notes for the visit by NDP government officials to TRW headquarters in Washington, D.C., in February 1992 state, "This is the first such MOU that TRW has signed with a Canadian province."

This MOU reflects the danger of further integrating the Canadian and Saskatchewan economies into the U.S. military-industrial complex in the wake of the FTA and NAFTA. The federal government granted the TRW team a multi-million-dollar contract to develop the Tactical Communications, Command and Control System for the Candu-3 project. This team, called the Computing Devices Company, involves both the Space Engineering Division (SED), based in the Research and Development Park at the University of Saskatchewan, and Information Systems Management (ISM), formerly Westbridge of Regina, based at the R&D Park at the University of Regina. A $20 million contract between the TRW and the SED was under negotiation in January 1992.

The briefing notes state that "smaller Saskatchewan firms and the universities" have been identified "as potential recipients of contracts and technology transfer benefits." They continue:

> Saskatchewan can gain significant benefits through a strong relationship with TRW. To enhance this relationship it is recommended that officials... include substantial opportunities for Saskatchewan industrial and research benefits in the planned AECL/TRW MOU. SaskPower's assistance and cooperation are key to this effort.

It is clear from these briefing notes that the TRW's involvement in Saskatchewan was an outcome of the AECL-SaskPower agreement on the Candu-3 and nuclear waste research. They refer to discussion with the TRW's vice president about "development opportunities in Saskatchewan, some of which will involve the AECL agreement."

The briefing notes also refer to the "AECL Strategic Alliance," saying "in recognition of Saskatchewan's pivotal role in developing the AECL/TRW relationship, both firms have expressed a strong desire to ensure Saskatchewan directly benefits from the opportunities emerging from their alliance." The two major areas identified are "simulation, process monitoring and control systems" for the Candu-3 and existing Candu reactors and "management and disposal systems for hazardous wastes, including nuclear." This involved working with Ada, a modern software engineered computer language developed for the U.S. military and required for most space and defence projects in Canada and NATO countries.

The AECL was already a member of the Saskatchewan Ada Association, which also included the SED, ISM, Saskatchewan Research Council and some other organizations. Both universities were *ex-officio* members and "are doing work on the study to review the curricula of major Canadian and U.S. universities to help introduce these technologies into the province." This may help explain why some faculties at the University of Regina were so eager to hone their relationship with the AECL.

The Saskatchewan Ada Association spearheaded a Saskatchewan Software Engineering Centre to help introduce this technology to the province. The briefing notes state:

> AECL and TRW have indicated that the proposed Saskatchewan Software Engineering Centre is complementary to this [Candu-3 and nuclear waste] effort.... Ratification of the AECL/SaskPower agreement is essential in capturing benefits in Saskatchewan.

These disturbing revelations confirm that the AECL-SaskPower MOU was tied into nuclear wastes all along. They also show that the AECL Candu-3 plans were interlocked with U.S. military corporations. The links to a nuclear weapons missile manufacturer like the TRW show how amoral the quest for economic diversification had become. It was going to be "jobs at any cost."[12]

The leaked briefing notes recommended "that the Minister... confirm Saskatchewan's commitment to the joint (TRW) MOU and... provide, if possible, some insight as to the province's position regarding the SaskPower agreement with AECL." In view of these revelations it became urgent to ensure that the newly elected Romanow NDP government went down a different road: one that promoted a sustainable and peaceful future here and abroad. Under growing pressure, the Saskatchewan NDP government did not sign the final agreement on the Candu-3, and the agreement with the TRW hopefully fell by the wayside. Though we felt impotent in the face of such immense continental, corporate and state power, our resistance to the nuclear nightmare was beginning to pay off.

Notes

1. ICUC's history can be located at <http://www.icucec.org/about.html> accessed June 2007.
2. This is documented in depth in Jim Harding (ed.), *Social Policy and Social Justice*, Wilfrid Laurier University Press, 1995, pp. 341–74.
3. See John W. Warnock, *Natural Resources and Government Revenues: Recent Trends in Saskatchewan*, June 16, 2005, available at <http://policyalternatives.ca/index.cfm?act=news&call=1124&do=article&pA=BB736455> accessed June 2007.
4. See "CIC Sells Remaining Cameco Shares," Crown Investments Corporation of Sask., Feb 15, 2002, available at <http://cicorp.sk.ca/cgi-bin/newsarchive/2002/04>, accessed June 2007.
5. See Harding, *Social Policy and Social Justice*, pp. 353–58, for more details.

6. See Harding, *Social Policy and Social Justice*, pp. 363.
7. See *The Commonwealth*, Vol. 51, No. 9, Nov. 1991, pp. 21–23, for all resolutions discussed in this chapter.
8. See "Energy report receives thumbs-up," Regina *Leader Post*, Nov. 6, 1991, p. A8.
9. See K. O'Connor, "Nuclear power in Sask.," Regina *Leader Post*, Nov. 18, 1991, p. A2.
10. TRW comes from the merger of Thompson-Ramo-Wooldridge, three companies in the U.S. aerospace and aircraft industry in the 1960s.
11. See "How to Sell a Nuclear Dump," *Harper's Magazine*, March 1992, p. 22. The Geo. Bush Jr. U.S. administration proceeded with the Yucca site soon after it took power in 2001.
12. See Jim Harding, "Jobs, But Not at Any Cost: Economic Growth and Ecological Decline," in Graham Riches and Gordon Ternowetsky (eds.), *Unemployment and Welfare*, Garamond, 1990, pp. 235–54.

chapter 14

DARK SIDE OF NUCLEAR POLITICS

After the uranium spill at Wollaston Lake in 1989, pressure was mounting for a new public inquiry. Much had changed since the Cluff Lake and Key Lake boards of inquiry (the CLBI and KLBI), and several new uranium mines were already under development. Though Saskatchewan NDP policy supported a uranium phase-out, the new Romanow government quickly agreed to hold a joint federal-provincial panel (JFPP) on uranium developments in the north. This proved to be "Roy's way" of addressing the conflict between party and government uranium policy.

In January 1992, the JFPP announced it would hold meetings to get public input on guidelines for the environmental impact studies (EISs) for these new mines. But these would involve only two of the five projects, the mines at Cigar Lake and McArthur River, as the other three mines already had EISs prepared without a public inquiry. Though this compromised the new inquiry, the IUC decided to participate in order to articulate the biases of past inquiries, hoping they might not be repeated.

Learning from Past Inquiries

On March 2, 1992, I presented to the JFPP on behalf of the International Uranium Congress (IUC). I suggested to the JFPP that before establishing guidelines for EISs it would be helpful to look back at the terms of reference and outcomes of past uranium inquiries. According to its final report in 1978, the scope of the CLBI was two-fold: to look at Amok's proposal for the Cluff Lake mine and to consider "the future expansion of the uranium industry in Saskatchewan." The KLBI had a much narrower, predetermining scope, publicly justified because the government had previously accepted the CLBI's recommendation that uranium-mining expansion be approved. I argued that the credibility of the overall inquiry process remained open to serious question as major lake drainage and building of canals was undertaken at Key Lake in 1978, before the CLBI had finished its report and months before the KLBI had been appointed.

It is noteworthy that Jack Messer, Minister of Mineral Resources at the time, who approved this drainage in June and October 1977, acted as campaign manager for the NDP in the 1991 election that returned the Romanow NDP to power, and soon after he was appointed president of the electrical utility, SaskPower. Messer exemplifies how a few "proponents" were recycled across institutions during the expansion of uranium mining and the battle over nuclear power in Saskatchewan. Another example is Roy Lloyd, who was the province's chief planner during the CLBI, then went to the SMDC, then to the Key Lake Mining Corporation, and finally to the KPMG, which did consulting for the AECL. Before his death, he also became associated with the Canadian Nuclear Association and the London-based Uranium Institute.

As detailed in chapter 4, the Saskatchewan public only became aware of these behind-the-scenes processes when Department of Environment memos were leaked in June 1979 and a court injunction to try to stop the work at the Key Lake mine had failed, in part because most of the work had already been done. The controversy returned to the courts after it was found that the Key Lake project was continuing even though its lease had expired in June 1981. The attorney general at the time, later to become premier, Roy Romanow, retroactively granted this lease in September 1981. Charges were later laid on the KLMC on behalf of Saskatchewan Environment for the failure of the Key Lake project to get approval from the *Water Resources Act*, and a small fine was levied in October 1981.

In other words, I pointed out to the JFPP that many of the same kinds of political maneuverings that had occurred around the federal environmental review of the Rafferty-Alameda dams under the Devine Tory government had occurred earlier with the uranium mines when the NDP was in power. But, at least in the case of the dams, the Environmental Assessment Panel challenged the province's actions by resigning after the Devine government refused to follow a court order putting an injunction on further work while the federal Panel was still sitting.[1] Saskatchewan's KLBI was far more compliant with the NDP government's disregard for the rule of law in the case of the Key Lake drainage.

Saskatchewan's corporate media mostly ignored these matters in the case of uranium mining, though not in the case of the Rafferty-Alameda dams. Was it because they bought into the uranium-boom forecasts, made by the Blakeney NDP government, which proved to be so erroneous? Was it because the environmental conditions facing Indigenous people in the north weren't seen to be as important as the environmental issues facing rural people?

I reminded the JFPP that all Indigenous groups at the time of the CLBI called for a moratorium on uranium mining, in part to settle land claims, and that two-thirds of the participants in the local hearings either opposed the expansion of uranium mining or supported a moratorium, but even so the CLBI accepted the proponents' arguments and recommended the go-ahead. This came as no surprise as content analysis of the CLBI transcripts, which I had undertaken, clearly established that all three CLBI commissioners had a pro-nuclear industry bias.

My recap to the JFPP also included the following: that cumulative uranium revenue to the province from 1977–91 had been only $200 million, not the $1.5 to $3 billion projected during the CLBI; that SaskPower's November 1991 *Energy Options Final Report* showed that energy conservation was far more advanced than the nuclear proponents wanted us to believe in the 1970s; and that the revolution in no-waste renewables and energy conservation and efficiency was continuing to undercut demand for nuclear power, as it had for the previous decade. Finally, I noted that, since these past uranium inquiries occurred, the 1987 United Nations (Brundtland) report on sustainable development explicitly acknowledged the serious environmental, technological and developmental problems with nuclear energy and made the call for renewable forms of energy to be a global priority.

Though the CLBI process and findings were seriously compromised, its terms of reference were fairly broad. The Board was to "review all available information on the probable environmental, health, safety, social and economic effects of the proposed uranium mine and mill...." That this was interpreted broadly in some cases, though not regarding Aboriginal rights, is shown by the inclusion in its final report of chapters on nuclear power and nuclear wastes, nuclear weapons proliferation and terrorism, and moral and ethical issues, which, of a total of ten chapters, comprised about one-third of the total text.

In retrospect, it is easy to see how the political economic bias of the times, something to which it is easy to fall victim, affected attitudes and decisions. The CLBI accepted common assumptions about energy supply and demand that favoured the nuclear proponents. Speaking about the broader scope of its inquiry, the final report said,

> The positive broader implications, too, have a local as well as a global scope, and include the benefit of ample energy in the form of electricity, not only to take the place of energy now being produced from certain fast-disappearing fossil fuels but to satisfy the ever-increasing new demands for energy, and in so doing to become one of the transitional sources of energy until ultimately renewable energy, particularly solar energy, now only in the rudimentary stages of development, becomes the chief form of energy.

This statement in the section on the Terms of Reference can be taken as a fundamental position of the Board. It shows that the CLBI made a number of assumptions: that electricity should replace demand for fossil fuels; that fossil fuels were quickly disappearing; and that there would be ever-increasing demand for energy. The sub-text was that uranium fuel and nuclear power would be the foundation for the coming "electrical society," to replace the "oil society." In retrospect the Board exposed its bias, since electricity is only a small percent of overall energy use, nuclear energy provides only 17% of this (only 14% in Canada) and can't replace most fossil fuel uses. While fossil fuels were becoming scarcer, new oil supplies (including offshore and heavy oil), which at escalating prices were profitable to develop, were being located. The Board also showed no appreciation of the technological revolution in energy conservation that was already underway or of its importance in averting the worst-case scenarios of global warming. And while it did show marginal interest in renewable energy, it falsely assumed that this was only at a rudimentary stage.

I proposed to the JFPP that the 1990s have clearly demonstrated that the CLBI's assumptions regarding uranium revenue, environmental protection, northern development and employment and global energy trends were seriously flawed. The initial case made for the expansion of uranium mining has been fundamentally eroded.

A Realistic Look at the Joint Panel's Terms of Reference

More similar to the CLBI in 1978 than the KLBI in 1981, the mandate of the JFPP in 1992 was broader than the specific mining projects under review. Its terms of reference included an historical re-examination and consideration of the cumulative effects of the uranium industry and mine decommissioning in the north. This suggested a whole series of vital questions from a more holistic, ecological perspective. The terms of reference, however, contained a clear bias that could work against this perspective. Though the second mandate of the JFPP indicated that some or all of these mining projects could be found to be either unacceptable or acceptable, the specific points in the first mandate were clearly slanted towards the mining company proponents, for example, points that stressed the opportunities, adequacies, mitigation and overall benefits of the uranium mines. A balanced approach would also focus on the inadequacies, costs, risks and alternative scenarios.

That the terms of reference included the possible rejection of any or all of these mines meant that the government sponsors recognized, at least in theory, that the CLBI no longer stood as a policy basis for justifying uranium mining. This meant that the JFPP could reconsider the adequacy of past assumptions, which tended to support the proponent. The more holistic potential of the JFPP was encouraging, though my concern about bias was buttressed by my awareness that this inquiry came belatedly, almost as a political afterthought. Since the public inquiries of 1978–80, new mines had not been well scrutinized, even though the CLBI suggested that "future projects may each require separate inquiries." The CLBI was adamant that a "separate assessment should be made of" phase II of the Cluff Lake mine, which never occurred.

Like the CLBI and KLBI, the JFPP began its work facing a wary public.

Its joint federal-provincial sponsorship may have been an attempt to avoid a public clash between the two levels of government, but a half-baked inquiry process on uranium mines already under development did not start out looking any fairer than the earlier uranium inquiries. And participants rightly asked whether there was a chance that any of these projects would be found unacceptable; or, if the inquiry were to recommend against a mine, whether the government would heed the recommendation. As we shall see, this skepticism proved to be well founded.

Recognizing that there were still opportunities to create more balance and a fairer, more independent review process, I made a number of recommendations. First, I suggested that the JFPP commission the environmental impact studies and sub-studies from bodies independent of the uranium and nuclear industry. Otherwise there was too much room for conflict of interest and errors of omission.

Second, I proposed that a more credible cost-benefit analysis be undertaken, arguing that there was a need to fully scrutinize the alleged and actual benefits from past uranium mining, and to place this squarely in a more accurate cost accounting context than had been the practice. (The phrase benefit-burden analysis is more comprehensive and encourages the analysis of distributive or social justice.) This would include, at the very least, the massive shortfall in projected revenues from the Cluff, Key and Rabbit Lake mines; the direct investments and indirect subsidies to the uranium industry in comparison to these returns; and the relatively low level of employment provided for northern Indigenous people. I argued that the revelations of the Romanow NDP government's Gass Report on the province's desperate fiscal situation,[2] which outlined the writing down of the value of the uranium mining company Cameco, when it made its public share offerings, should also be explored in depth.

Third, I raised the long-ignored matter of Aboriginal rights, suggesting, in light of constitutional renewal underway in Canada that the JFPP should consider the implications of "inherent" Aboriginal rights and moves towards self-government for the future of this industry. Such rights had already been affirmed in a federal court, in 1979, arising out of Inuit concerns about uranium exploration in the Keewatin region of the eastern Arctic. In the Sparrow case in 1990, the Supreme Court of Canada stressed the primacy of Aboriginal rights over commercial rights in the west coast fishery.[3] It may be that inherent Aboriginal rights will someday alter the ability of uranium companies to exploit non-renewable mineral resources. At the very least, revenue sharing may be required as a form of compensation, and I suggested looking at the experience of Australia's Northern Territories. In Australia, Aboriginal Land Councils have a four-year veto on uranium exploration, which gives them some bargaining power not available to Canadian First

Nations or Métis. Later we shall see that the JFPP did follow up on this, as it did propose revenue sharing.

Fourth, I encouraged the Panel to carefully study the questions posed by the Inuit coalition preparing for the FEARO inquiry into the Kiggavik uranium mine proposed near Baker Lake, N.W.T. I suggested the guidelines on social, environmental and economic impacts should at least be as comprehensive and stringent as those required by that environmental review panel.

Fifth, I raised the matter of environmental health protection and, as part of the assessment of the history and cumulative impacts, I encouraged the JFPP to look at the actual record of spills and tailings management problems at existing mines and to closely compare this with what was guaranteed by uranium companies about "state of the art" uranium mining.

Sixth, since there has never been a baseline health study completed for the northern areas where uranium mining is occurring, I argued such a study must be completed before any consideration is given to further expansion of uranium mining in the north. At the time, it was thirteen years since such a study was first recommended. It was particularly vital for extensive baseline studies to be done in the Wollaston Lake area, where the most concentrated activity of the uranium industry is occurring. I also suggested the JFPP ask for projections of all northern hemispheric releases of radon gas from the tailings from these proposed new mines, for the extremely long-term duration of radiation release. The importance of this global-ecological understanding of radiation and health had become more apparent since the Chernobyl accident in 1986.

Finally, I pointed out that past social impact analyses had been superficial and incomplete. This time there must be extensive, independent studies undertaken that fully recognize the cross-cultural dimensions and account for the role of the renewable and harvesting economy in the Wollaston Lake area and the wider north and for the steady evolution of Aboriginal self-government.

After the disappointing experience of the CLBI and KLBI these were the major areas that needed attention. As part of this, I encouraged the JFPP to consider the implications of emerging environmental law for government regulation of this industry; how market forces are beginning to shape innovations in energy technologies; and, finally, how the uranium and nuclear reactor industries are moving towards an integrated strategy in this province to try to ward off these challenges to their survival. Later, during its Technical Session on Socio-Economics, held in May 1993, I followed up and gave the JFPP 200 pages of documentation on many of these matters.[4]

Those of us closely watching this new inquiry would be able to tell whether the JFPP seriously considered the implications of their broad mandate by looking at these matters or succumbed to being another nuclear white-wash. By then, after more than a decade of dealing with government-appointed apologist bodies, I was deeply skeptical about what lay in store. However, the JFPP was to hear similar skeptical views across Saskatchewan and began to take seriously some non-nuclear concerns. Later, it came as a pleasant shock that the JFPP recommended against some of the uranium projects. The shoe was soon going to be on the other foot.

NDP Pro-Uranium Again: Hollow Victory in Global Context

In November 1992, while the JFPP was still undertaking its hearings, the NDP convention reversed the party's position back to supporting uranium mining. The vote was close and the split in the party was deep, but the pro-nuclear forces won the day. Until this convention it looked like non-nuclear sentiment was on the upswing. Seeming to follow the lead of the Energy Options report, in March 1992, Premier Romanow announced the creation of the Saskatchewan Energy Conservation and Development Institute (SECDI), which only lasted until 1996. In spite of a massive business lobby, the NDP government took the lead from its party and caucuses and cancelled the SaskPower-AECL MOU it had inherited from the Devine Tory government. There was strong lobbying on Cabinet from both provincial and federal caucuses to follow party policy on this matter.[5] There were, however, indications that the Romanow government was going to embrace pro-nuclear policies whatever the party decided. Just after the December 1992 convention, the NDP government announced it had signed a new MOU with the AECL. While this MOU made no commitment to a Candu and notably rejected nuclear waste storage, it agreed to move the Candu-3 design work to Saskatoon and to explore the uses of the Slowpoke reactor. It looked like Romanow and his cabinet had successfully stick-handled through the grassroots of the party.

The 555 to 437 vote that reversed uranium policy pleased the uranium industry and its supporters in business, the university and the media. But it marked a setback for those working for a non-nuclear, sustainable future. And it wasn't going to settle anything in the long run, since this long-standing and complicated controversy promised to remain with us — and our offspring's offspring — for as long as the half-life of the industry's radioactive by-products.

The NDP flip-flops on this matter have become common. In the 1978 provincial election the NDP promoted the expansion of uranium mining as the means to Saskatchewan becoming a "have" province. It was to be the province's golden egg, but costs were soon to outweigh benefits. In opposition from 1983 to 1991, the NDP held to a policy of phasing out uranium mining. Then in 1992 the governing NDP supported the further expansion of uranium mining, but this time to try to avert further economic decline. What was to be a path to billions of dollars of provincial revenue became nothing more than a holding action.[6]

Convention Process Flawed or Rigged?

A close look at the policy-making processes at the 1992 NDP convention raises questions about whether there was fair and open decision-making. The adopted pro-uranium policy cleverly made further uranium mine expansion conditional on the environmental review process, the JFPP, to which the Romanow government had already committed. Based on the outcomes of past inquiries, this seemed to carry few political risks.

Government and party officials twice (in 1991 and 1992) refused a province-wide educational process as a prelude to any vote. The proponents of uranium mining were apparently leery of freedom of inquiry, after what happened in 1983 when the party adopted a uranium phase-out policy. Without an open process the industry had an advantage through its million dollar promotions and backroom lobbying, and the appeal of simplistic clichés about economic growth and jobs in a tough economic climate.

Most vital, Premier Romanow implied publicly before the convention that the government's economic development strategy involved going ahead not only with uranium mining, but with a new MOU with the AECL. (Though it was not widely reported, his press release announcing the formation of the SECDI indicated he wanted to renegotiate the AECL deal.)[7] These were, he said, government policies regardless of what the party decided, siding with uranium industry backers in the government, who were appealing to the slight 1992 convention victory as some sort of justification for government policy. Romanow's end-of-convention appeal to "come together" now that the party had "decided" on this policy certainly didn't ring true to anyone who knew the messy, manipulative background.

Politics is fickle, and it is likely that the Premier's statement about going ahead regardless of party policy influenced some delegates to vote for a reversal of the policy, to try to avert the party and government appearing to be at loggerheads. Such is self-fulfilling power politics, with party loyalty and public optics, not policy principles and analyses, increasingly driving partisan decisions. Immense pressure was put on some delegates and caucus members to support the government position. It only took a shift of a mere sixty votes out of nearly a thousand delegates for this policy reversal to occur. Delegates who were also executive assistants in the employ of the new NDP government could have had a major impact on this policy reversal. It is perhaps understandable why some citizens believe that in a corporate-dominated society, "democracy" is not about public control of decision-making but a hurdle that the corporations must overcome.

The vote might have gone the other way if delegates had known before the convention what they had learned just a few weeks later: that the reformulated MOU with the AECL not only included design work on the proposed Candu-3 nuclear reactor but the possibility of constructing a Slowpoke reactor business in the province as well.[8] One can only speculate that the announcement of this deal was purposely delayed until after the convention.

The vote that reversed party uranium policy reflected procedural and semantic maneuvering more than an assessment of energy policy options. There had been much pro-industry strategizing since the 1991 post-election-victory convention, when the attempt to support the AECL deal made by the Devine government was defeated soundly and the Steelworkers failed to get a pro-uranium mining resolution to the floor for a vote. The pro-nuclear lobbyists within the government bureaucracy had a distinct organizational and resource advantage over the grassroots non-nuclear party members. And judging by the way the 1992 convention debate was choreographed, they had clearly learned from their 1991 defeat. The resolution that ultimately carried was the fourth to come before the convention's panel on the economy, whereas the resolution reaffirming party policy (the phase-out of uranium mining) was far down the list, the 38th resolution. It was probably no accident that all the pro-uranium mining resolutions were at the start of the panel, while those affirming non-nuclear options were near the end. Those in high government and party places were now covering all angles.

But there was also ideological manipulation. The wording of the pro-industry resolution that became the focus of the controversy at the convention was clever, even cagey. It talked of the desperate need for jobs, private investment and revenue, and it stressed the existence of the JFPP. It asserted that there was a "majority of northern people who support the prospect of further uranium mining," something that would be very hard to substantiate. Direct research done by myself suggested something very different.[9] Rather than explicitly affirming a pro-uranium position, the resolution ended with the twist that the government "approve any new uranium mine developments only after it receives a report" from the JFPP "that all environmental protection, health and safety standards and northern economic development objectives have been met."

Officials in the uranium industry couldn't have better designed the resolution. It stressed that the new mines would bring "private investments of $520 million during the 1992 to 1996 period, the employment of an additional 1240 people by 1996 accounting for $350 million in wages, salaries and benefits." It emphasized that this would involve "very little investment in additional infrastructure by the provincial government." The era of expanded uranium mining under the new Romanow NDP government wasn't to be marked by joint ventures or public enterprise, as it was under the Blakeney NDP government, but old-fashioned free market capitalist growth. The industry didn't even have to threaten a capital strike, or withdrawal of jobs, as is so often done when politicians talk of environmental regulation or shutting down polluting industries.

When the uranium industry expanded in the late 1970s it talked of 2200 jobs, only half of which materialized. The promise of another 1240 jobs by 1996 was therefore to be taken with a big grain of salt. Even if these jobs had all miraculously materialized, which they haven't, the cost per job — even discounting all the past investment in the facilities — would have been over $400,000.

In retrospect, the job promises made in 1992 appear inflated and manipulative. In 2002 the NDP government reported 4,300 jobs from uranium mining, but this included "indirect jobs," which leaves lots of room for massaging statistics. But even taking this figure, at $3.2 billion invested in the industry to that point, the cost was now $750,000 per job, hardly a sustainable economy. In a 2006 series on the uranium industry, the Saskatoon *Star Phoenix* concluded that there were 2,159 people directly employed in the uranium industry, including contractors, so even with the new mines operating, the jobs remain at about half of what was projected. This multi-national megaproject approach to job creation can never provide a viable or sustainable economy, in the north or anywhere else.

The resolution was intended to reassure people with concerns about uranium mining that someone else (the JFPP) would look after these matters. It provided a diversion for delegates who didn't want to tackle the fundamental questions about environment, development and peace. People could vote to reconcile the party position with what the Romanow NDP government was going to do anyway and yet somehow believe someone else was going to deal with the outstanding issues. They were being given the opportunity to have their political cake and eat it too.

Anyone closely watching the uranium inquiries since the 1970s knew the appointments, terms of reference and processes were all biased towards the nuclear proponents. And the JFPP continued with the tradition of not looking at Saskatchewan uranium's end-uses around the globe, which meant that many of the most vital questions were ruled out. To its credit, however, as we shall discuss below, the JFPP, chaired by University of Regina's Dr. Don Lee, still raised this matter in its final report.

Considering the procedural and semantic twists and immense pressure on party members to be compliant with the Cabinet and Premier, it was perhaps a sign of things to come that nearly one-half of the delegates did not succumb. Even after the reversal of policy there was a much more informed base of support for non-nuclear energy and development options than under the Blakeney NDP government in the 1970s. The backers of the pro-nuclear policy reversal clearly had the upper political hand but at the cost of ethical, intellectual and policy credibility.

The NDP flip-flops from 1978 to 1983 to 1992 show an underlying confusion about how to create a sustainable society; it also suggests that party policy may be fairly irrelevant, except when it gets in the way of government and corporate policy, in which case you change it. If (when) the Calvert NDP that followed on the heels of Romanow loses after its second or third term, will the party return to a non-nuclear policy? And then what happens if the NDP is again re-elected?

An interesting comparison can be made with Ontario when the Bob Rae NDP government (1990–95) pursued a non-nuclear energy policy and looked at energy efficiency and conservation. Once elected, it did not collapse and reverse its policy to satisfy the nuclear business lobby. It reined in the expansionist and inflationary Ontario Hydro with its plans to construct additional nuclear power plants. It even brought in Maurice Strong, the Canadian member of the U.N.'s Brundtland Commission, to oversee its review of Ontario Hydro. Though the Rae government went down to defeat after one term, largely because of its disastrous "social contract," these energy initiatives were the first governmental attempt to challenge Ontario's blindly taken nuclear path.

What the Romanow NDP government did is more akin to what the Peterson Ontario Liberal government did. After being elected in 1985 with a mandate to stop the expansion of nuclear power, Peterson quickly flipped and continued the multi-billion dollar construction of the Darlington reactor. The Ontario taxpayers are still paying for this lack of political backbone.

The nearly even split at the Saskatchewan 1992 NDP convention and the significant paradigm differences that this reflected likely contributed to some political realignment. Certainly the "nuclear issue" played a major role in the political realignment that gave rise to the Greens in Europe, and after 1998, a Saskatchewan Green Party (originally called the New Green Alliance) began running candidates. Its percentage of the vote actually went down in 2003 compared to 1999. The group needs to continue to develop a more in-depth, concrete Saskatchewan-based integration of ecology, peace, democracy and social justice. Realistically, without proportional representation, the provincial Greens may remain more a "moral" than electoral force.

As premier, Romanow was successful in getting the NDP closer to being a liberal, centrist party along the lines of Tony Blair's Labour Party, though he may have been undermining the coalition that brought him to power. Deep grassroots disappointment existed with his government's continuation of many trends started under the Devine government, including the drift towards nuclear power and talk of further privatization. Some of those who worked so hard to oust Devine were a lot less committed to work for the NDP in the future. The Romanow NDP nearly lost the 1999 provincial election to the right-wing, Alliance-like, Saskatchewan Party, having to form a coalition with the small Liberal caucus to hold onto power.

Nor was the federal NDP pleased with the Saskatchewan NDP policy turn-about. The federal NDP caucus voted 90% against the Saskatchewan NDP keeping the SaskPower-AECL MOU.[10] And the way the policy reversal was brought about did nothing to affirm the role that federal leader Audrey McLaughlin had envisioned for the NDP at the November 1992 Saskatchewan NDP convention, as the "conscience of the nation."

Consent Can Be Manufactured

The uranium industry carefully crafted media support to cement this shift in NDP uranium policy. One of the most vocal and crass nuclear supporters has been Regina's *Leader Post* business columnist Bruce Johnson, who, in his February 29, 1992, column, made it clear he was in total agreement with the SaskPower-AECL MOU on a Candu-3 for Saskatchewan. In his November 7, 1992, column before the NDP convention, he contradicted the research on the health risks of uranium mining, by stating "uranium mining is no more risky than other mining activities, while radiation exposure risk for most miners is no greater than that for office workers." Johnson had only to visit his local doctor's office to get his research straight. Compare his assertion to what the Canadian Cancer Society states about lung cancer in their widely available (December 1989) pamphlet *Cancer Facts for Men:* "Employees exposed to asbestos, chrome salts, nickel refining, coal tar products and radio-active uranium have been found to have above average risks of developing the disease." In his blind support for economic growth regardless of consequences, Johnson labels those who support safe non-nuclear options as "anti-nuclear zealots," "anti-development Luddites" and the like. Since epidemiological studies continue to show a much higher than average risk of lung cancer among uranium miners, the pro-nukes are willing to sacrifice not only truth but workers' health to keep the nuclear bandwagon afloat.

The *Leader Post* column by Murray Mandryk was thankfully less partisan. Before the 1992 convention, in his November 7 column, he claimed the "government forces" were trying to ensure the continuation of the uranium industry for strictly economic reasons. He, however, still asserted that support for this resolution would "be a victory for the job-creating pragmatists." He showed no interest in assessing either the short- or long-term costs (including environmental) of jobs in this industry, but rather commented, "any opposition to jobs... doesn't seem very practical in this day and age." He was, however, probably right that "This debate isn't about nuclear weapons or nuclear reactors to most. It's about the future of the uranium mining. It's about jobs."

To some extent this shows that the CNA and AECL strategy of emphasizing economic development and promoting nuclear amnesia was having an effect.

Comparative facts, however, just don't support the uranium industry as a job-creating strategy. One thousand jobs by 1992, a small minority of which go to Indigenous northerners, isn't very impressive when you consider the billion dollars of investment that has gone into this industry since the late 1970s. Far less investment into other energy and development options could have created far more jobs. Unfortunately, comparative facts are rarely considered in a political climate of globalization, where there is such pressure for investment and jobs at any cost and there is such a total lack of imagination or courage among politicians of all stripes.

Northern NDP MLA Keith Goulet argued at the 1992 convention that northerners aren't dying of radiation but of unemployment. While the rhetoric may be appealing, this misses the point that uranium mining contributes to structural underdevelopment by diverting vast amounts of wealth into capital-intensiveness, which is then not available for sustainable labour-intensive industries. And this cannot overcome northern unemployment. While Goulet's righteous plea about northern poverty was probably good for his career as a cabinet minister, it ignored the fundamental questions about the end uses and wastes generated from uranium mining. This "means justifies the end" argument is basically the same as that made on behalf of un-ecological ranching, mining or agribusiness in Central and South America, as a justification for the destruction of rain forests and of Indigenous culture and local economies that depend on them. The world's ecological and cultural diversity continues to be sacrificed by such appeals to short-run, trickle-down benefits from corporate economic growth.

Nuclear Foot in Saskatchewan Door

While a reversal of party policy was accomplished by narrowing the focus to short-term and self-interested market economics, the NDP was now left with a serious contradiction, which was to continue to plague it. The Saskatchewan NDP was again officially in support of shipping uranium fuel to other regions and countries of the world for use in nuclear reactors while at the same time increasing the risk of nuclear weapons proliferation. Yet, at least for the time being, it remained opposed to constructing a reactor or storing nuclear reactor wastes in the province. The NDP government and a slight majority of the party delegates in 1992 were willing to seek questionable short-term benefits from uranium expansion, with the devastating long-term risks from the spread of the global nuclear system being assumed by others.

This amorality cannot be swept away with clever words. The nuclear industry will continue to use this contradiction to try to construct reactors and nuclear waste storage facilities in Saskatchewan. The 1992 policy reversal on uranium mining clearly gave the nuclear reactor and waste industry a bigger foot in the province's door. Within a few weeks of the NDP policy reversal, the AECL announced they were moving 140 jobs from Ontario to Saskatoon, as part of the new MOU. In announcing the agreement on December 21, 1992, Federal Minister Jake Epp said, "I'm delighted to welcome Saskatchewan into Canada's nuclear family...."[11] Soon after, the president of the partly provincially owned uranium company, Cameco, was again in the press advocating that Saskatchewan "rent" its uranium to nuclear customers, with the understanding that Saskatchewan would take back nuclear reactor wastes.

The resolutions for the 1992 NDP convention were entitled "New Hope—New Direction," but in the light of what uranium leaves in its wake and given the existence of practical energy alternatives with immense social and economic benefits, this reversal of party policy maintains an old direction, without hope. Put in the global ecological, human development and ethical context, the 1992 reversal of NDP policy on uranium mining was a hollow victory indeed.

Red Light or Green Light?

The new NDP policy made new uranium mines conditional on approval by the province's mandated environmental review process. The JFPP submitted its first report in October 1993. In spite of clear pro-nuclear bias in its terms of reference, the JFPP forcefully rejected one mine and called for a five-year postponement of another — sending shock waves through the corporate and bureaucratic world.

Finally, after years of presentations at public inquiries and being mostly ignored, non-nuclear activists felt vindicated. Issues of cumulative environmental and social impact and economic underdevelopment were being taken seriously. Our renewed confidence in the public review process was, however, short-lived. It soon became clear that not only was party policy of no real consequence to the governing Romanow NDP, but neither was its own environmental review process. By then the Saskatchewan NDP had earned a nation-wide reputation for flip-flopping on uranium mining. There is probably no issue over which it has squandered so much of its political, even moral, capital, and it was about to risk losing the very little it had left.

Industry, government and environmental groups have all become used to the rubber-stamping of uranium mines by reviews, so it came as no surprise that the JFPP conditionally approved the expansion of the Cluff Lake mine because it deemed the risks "incremental to those already in existence" and suggested they "could be reduced to acceptable limits provided certain conditions are met." However, to everyone's surprise, the JFPP found the Midwest project completely unacceptable because "the benefits that could be obtained are insufficient to balance the potential risks." It also called for at least a five-year postponement of the McLean Lake project, which could only proceed under stringent conditions.

The rejection of the Midwest mine was comprehensive. The JFPP found that risks greatly outweighed benefits, mining methods were unacceptable and the high-grade uranium, arsenic and nickel posed serious problems in underground spaces. The company had left more than 600 exploration holes uncapped, posing a threat to workers' health. Contaminated effluent also posed an environmental threat and uncertainties remained about the disposal of uranium tailings. Finally, there were concerns regarding the cumulative effects of this mine and others, concentrated in the eco-region near Wollaston Lake, and the risks posed to the public by the transport of the high-grade ore on Saskatchewan's highways.

At the end of the contentious 1992 NDP convention, where the party barely reversed its position, Premier Romanow made the unambiguous promise, "If the panel says red light, there is no further mining in Saskatchewan. I can't see any government of any stripe saying we're not going to pay attention to that. I think that's the end of it."[12] However, once the JFPP reported to the NDP government, Romanow gave clear signals that the previous guarantee to respect party policy and the review process was not on. NDP Environment Minister Bernie Wiens broke with the agreed review process, saying he would take further submissions until the end of November. This opened the door for the uranium industry and its business allies to fervently lobby the government to not follow the JFPP's recommendations.

They were obliging, with one Saskatoon business criticizing the JFPP for listening to non-nuclear groups, as though its scientists could not assess arguments and information for themselves. The call for further briefs was widely criticized as two-faced and undemocratic, with the chief of the Prince Albert Tribal Council expressing concern that this decision was a way to try "to find more excuses to disregard the panel's recommendations, particularly those related to the cancellation or delay of the mining development."[13]

The 1993 federal election showed how disenchanted Canadians were becoming with the sincerity of politicians, and, ironically, this was one major reason for the quick growth of the Reform-Alliance-(non-Progressive)-Conservative party. (While alienation from government was a populist political tool to get power to move Canada further towards being an outright branch-plant of the waning American empire, we clearly won't see more political democracy under the Harper Conservatives.) And the handling of the uranium issue was to be a serious test of whether or not, once in power, the Saskatchewan NDP would provide any fundamental alternative to the cynical rule of the old-line parties. If the NDP government ignores its own review process and the Premier so easily goes back on his word on such a vital matter, how can anyone believe that the NDP is any more committed than the Liberals or Conservatives to public participation and environmental assessment?

The JFPP stated that failure to accept its recommendations contrary to the promise by the Premier would damage the credibility of the review process:

> The public perception is that recommendations made by the present panel may also not be acted upon by the government. This would defeat the intent of the review process and negate the considerable efforts made by the panel, members of the public, proponents and government departments to conclude a full and fair review.

While in opposition the NDP was very critical and a bit self-righteous about the Devine government's disregard for the federal review of the Rafferty-Alameda dams. Yet this 1991–93 review of uranium mining was not only federally mandated but was done jointly under provincial laws. Disregarding its own process, the NDP deservedly looked even worse than the provincial Tories in this regard.

Also, if the NDP ignored its own environmental review, it was hard to see how the JFPP could complete in good conscience its report on the remaining two mines, at Cigar Lake and McArthur River. The wider public would be discouraged about participating in public hearings, which would again be seen as a sham.

End Uses Finally Out of the Nuclear Closet

The JFPP was hamstrung from the start by terms of reference that ignored uranium's end uses. Nevertheless, JFPP members were convinced by their extensive investigations that they had to draw this matter to the attention of the provincial and federal governments. They wrote:

> There is no process whereby exported Canadian uranium can be separated from uranium derived from other sources. Therefore no proven method exists for preventing incorporation of Canadian uranium into military applications.... Current Canadian limitations on end uses of uranium provide no reassurance to the public that Canadian uranium is used solely for non-military applications by purchasers.

About the Midwest proponent, the JFPP stated: "Specific proponents, such as Cogema, are wholly owned subsidiaries of foreign governments heavily involved in military weapons research, fabrication and testing."

This view is not some aberration; the U.N.'s Brundtland Commission also raised a similar concern about the so-called "peaceful atom" fueling nuclear weapons proliferation. But the JFPP saying it clearly, in the context of widespread nuclear amnesia in Saskatchewan, was a vindication for what non-nuclear activists had been saying for over a decade.

At the time, Cogema had concentrated its control over the McLean Lake project as well as the projects at Midwest and Cluff Lake. Cogema, which along with Cameco was among the largest uranium/nuclear corporations on the planet, used its significant corporate clout to get around the JFPP's recommendations. It brought a revised proposal to convince the NDP government to reject the recommendation of a five-year postponement of its lucrative McLean project — demonstrating further that the environmental review process is respected by such multinationals only if it's onside with corporate objectives.

The JFPP recommended the postponement of the McLean Lake project for two primary reasons: first, to allow more time to evaluate the proposed (pervious surround) method of tailings management; and second, to allow time to ensure 80% northern employment and to negotiate revenue sharing with local communities. It wrote: "Thus, most northerners receive little, if any, benefit from the uranium mining industry because the economic system of the region fails to redistribute wealth." Finally, such influential people as the JFPP members acknowledged that there were enough serious flaws in the rhetoric about northern benefits to warrant a serious social impact reassessment.

The JFPP noted that standardized, quality information to allow assessment of the cumulative effects of mining has not been collected. Regarding McLean Lake, the members said: "It is not a question whether or not there will be cumulative environmental impacts, but of their magnitude." They were particularly concerned about the implications for the Indigenous harvesting economy:

> Cumulative impacts on that portion of the Athabasca Basin west of Wollaston Lake and south of Hatchet Lake might be considerable.... the overall effect of the operations, with the possibilities of interconnecting roads and power lines, would be widespread.... the entire area might become unproductive for traditional hunting, fishing and gathering activities.

They were finally saying what some northern Indigenous people had suspected, especially since the uranium spill at Wollaston Lake in 1989, but were too afraid to say for fear of political and economic repercussions.

The JFPP also agreed there were still no baseline health or epidemiological studies by which the impact of uranium mining could be fully assessed. Such studies had been recommended fifteen years beforehand, back in 1978 by the Cluff Lake Inquiry. Almost thirty years later, in 2007, there is still no baseline data, which is the first step in any credible social or health impact research. So the industry continues to be allowed to operate in the dark without any fundamental ecological or legal accountability. Like nuclear waste buildup, worker and community health has been put so far back on the back burner that it has fallen off the stove.

The choice couldn't have been clearer for the Romanow government. Either the NDP government of the day respected its review panel and party policy and made health, environmental protection and northern justice its priority; or it abandoned these for political economic expediency. Which was it to be? Would an NDP government finally stand up on the uranium issue?

Not surprisingly, uranium mining again became a burning issue at the November 1993 party convention. Some of us leafleted the convention to be sure delegates were confronted with the glowing contradictions. A resolution was carried, reaffirming the 1992 party policy, which meant the NDP cabinet should uphold the decisions of its review process, the JFPP. On a cold, clear wintry morning early in 1994, I waited with an activist colleague outside the pressroom of the legislative buildings for the NDP cabinet's decision. After years of intense involvement in the public inquiry processes I remained pessimistic.

In the end, industry — not the review process, the NDP or the north — prevailed. All mines got the go ahead, with minor conditions. It was to be business as usual. In the crunch the Romanow government turned its back on the party policy it had crafted. For all practical purposes, party democracy, as well as public participation in environmental protection, was dead in the province. The lines between the NDP and other parties on environmental credibility were now irrevocably blurred.

Romanow's time as premier was to become even more difficult. He barely scraped through the 1999 provincial election and had to form a coalition with the few Liberal MLAs to maintain power. He resigned mid-term. Later he headed up the federal commission on Medicare, where he rehabilitated his image as a progressive politician. The commission, however, did not look at the long-term health implications of his past government's uranium expansion policies. Unfortunately, Saskatchewan may yet become better known for its major role in global uranium contamination than for being the original home of Medicare.

Notes

1. See Canada Federal Environmental Assessment Review Agency, Report on the Environmental Assessment Panel, Rafferty-Alameda Project, 1991.
2. See *Final Report*, Sask. Financial Management Review Commission, Regina, Feb. 1992, p. 39.
3. See Section D8, in "Uranium and Public Policy," in *Uranium: A Discussion Guide*, National Film Board, 1991.
4. See Jim Harding, "Critically Assessing the Impacts of Uranium Mining on Aboriginal People in Canada," Socio-Economics Technical Session, Prairie Justice Research, University of Regina, Saskatoon, May 7, 1993.

5. See D. Eisler, "Internal politics led to conversion," Regina *Leader Post*, February 27, 1992.
6. In 2006 the government reported uranium sales of over $600 million, yet all taxes and royalties came to only $43 million. Economic Analysis, Saskatchewan Industry and Resources, June 29, 2007, personal communication.
7. "Saskatchewan's Electrical Energy Needs: A Major Economic Development Opportunity," *News Conference Statement*, Saskatoon, March 11, 1992, p. 7.
8. See *MOU Between Gov't of Sask. and AECL*, Regina, Dec. 21, 1992, p. 2.
9. See Jim Harding, *Impact Assessment Bulletin*.
10. See Dale Eisler, "Internal politics led to conversion," Regina *Leader Post*, Feb. 27, 1992.
11. "New AECL deal to bring 140 jobs to Saskatoon," News Release, Dec. 21, 1992, p. 1.
12. Reported in Regina *Leader Post*, Nov. 4, 1993, A6.
13. Reported in Regina *Leader Post*, Oct. 30, 1993, B1.

chapter 15

GLOBAL EDUCATION

I've been an educator most of my adult life, teaching in several universities in three provinces since the mid-1960s and find abhorrent the extent to which nuclear industry propaganda has penetrated our schools. By the early 1990s, industry material was being used in Saskatchewan government information kits and referenced as curriculum resources. Nuclear industry speakers, videos and pamphlets flooded our classrooms. *Uranium Today*, the industry's promotional trailer, a sort of travelling road show launched by a pro-nuclear lieutenant governor, went to schools giving "science class" tours. The International Uranium Congress (IUC) tried to counter-balance this bias with a low-cost kit, "Resources for a Non-Nuclear Future," which went to interested school libraries and science teachers, but it was no match for the multi-million-dollar resourcing of the industry. My eighty-year-old father, Bill Harding, single-handedly did most of the distributing of these resources across schools in the province. His death in early 1993 left a huge vacuum in my life as well as in the Saskatchewan non-nuclear movement.

The nuclear propagandists had to be flushed out from behind their curtain of deceit, and in February 1992, I was given the opportunity to debate an AECL spokesperson, Dr. George Sparks, at the Prince Albert Area Teacher's Convention. I did my homework on the gathering Candu crisis in Ontario (which I discuss in the next chapter) and was surprised to learn how ill-informed Dr. Sparks was — seeming to have bought AECL's simplifications, hook, line and sinker. The information I disseminated was being heard by many teachers in Saskatchewan for the first time, which made it more compelling. The AECL's vulnerability in the face of open adult education was becoming more apparent.

Establishing an Educational Framework

The basic understanding of the underlying issues, assumptions and arguments required from an educational viewpoint is difficult to attain in an adversarial context. According to our research in the Uranium Inquiries Project at the University of Regina, nuclear proponents and opponents don't only disagree on the specific issues of uranium mining and nuclear energy, such as radioactive waste, reactor accidents, nuclear weapons proliferation, they also differ in their basic worldview.[1] This quintessential difference of perspective is key to the confusion often experienced by the undecided citizen, so I wanted to look at the nuclear debate in more educational than adversarial terms. When we learned that nuclear industry spokespeople were presenting at provincial teachers' meetings, we contacted organizers and insisted that a public educational event should show better balance in viewpoints. After some negotiation, I was invited to present to teachers at the 1993 Regina Showcase conference.

In my talk to about a hundred teachers during the February 22–26 conference, I suggested that such a controversial issue has to be approached from a multidisciplinary orientation because overspecialization leads to rigidity, myopia and denial. The methodologies used in global education are extremely helpful, and an article, "Teaching Controversial Issues," by Pat Clarke, published in the *Green Teacher*, a magazine I strongly recommend to teachers, is particularly to the point.[2] Clarke's framework overcomes several obstacles faced by global educators when they teach controversial issues. These obstacles include, "a concern for the influence of a teacher's own biases, a fear of becoming a lightning rod for controversy oneself… and a lack of confidence because of unfamiliarity with an issue." Clearly, in the case of the nuclear issue the matter of bias, including huge institutional biases, must be addressed head on.

Clarke proposes what she calls a "de-mystification strategy" for approaching such controversial issues. This de-mystification involves four basic questions: 1. What is the issue about? 2. What are the arguments? 3. What is assumed? 4. How are the arguments manipulated? In my talk at the Showcase conference, I applied these questions to the nuclear controversy in the hope that this would encourage more educators to take personal and professional responsibility to balance this debate in their classrooms.

Clarke points out that values, information and concepts are all at play when defining an issue. This is certainly true in the nuclear debate, but it takes a lot of critical examination to disentangle and clarify these aspects. When assessing the arguments involved in an issue, Clarke notes that there is the need to critically analyze the claims and the nature of the support for such claims to be able to judge the validity of a position. (This, in a nutshell, is what the ideals of liberal education are supposed to be about.) And certainly very different claims are being made by nuclear proponents and opponents. Clarke also contrasts what she calls moral and prudential criteria for making judgments, noting that, "Prudential criteria are those that are concerned mainly with how someone or their group will be affected." What this means is that claims about nuclear power cannot be analyzed solely in terms of information, but must also be judged in terms of how they are connected to corporate or political self-interest or the public interest.

Regardless of the degree of logical support given either side of a controversial issue, proponents and opponents alike can easily get trapped in circular self-referential (or tautological) ideological arguments. Within Clarke's framework, these arguments are based on assumptions that are "taken as self-evident." They can also be seen as the first principles within the argument and must therefore be assessed in terms of their concrete consequences. In the case of nuclear power and nuclear weapons, the examination of these consequences raise direct questions regarding energy policy and environmental impacts, as well as questions regarding human rights, social justice and democracy.

Clarke also poses the question of who is articulating the argument and its underlying assumptions and principles. Who are the voices behind the controversy? This is timely to the nuclear controversy, as proponents often claim to have special knowledge not held by "outsider" critics. Clarke acknowledges that "insiders may have particular information and interests which could give an argument a certain shape or orientation" but asserts that "an argument can best be tested by hearing views of both insiders and outsiders." As I show below, the education system in Saskatchewan has been far more willing to hear voices primarily from within the uranium and nuclear industries.

Finally, there's the need to look at how controversial arguments can be manipulated, which involves assessing how information and misinformation are used to influence opinion. Clarke points out: "To determine how an argument is being manipulated students must first determine who is involved and what are their particular interests in the issue. What is the rationalization for their position? What are the reasons for taking the position they advance?" In our particular case, just why have the CNA, AECL and Saskatchewan Mining Association (SMA) placed their resources in Saskatchewan schools?

This approach helps show how information is selected or ignored and why errors of omission as well as commission are made. It puts the nuclear controversy into the context of, as Pat Clarke puts it, how "parties involved are acting in self-interest and use information only to support that interest." This requires examination of the role of the corporate media, how the "media can engage in argument manipulation" and contributes to the acceptance of a particular version of reality and "the truth."

In the nuclear controversy there are a plethora of issues, arguments and assumptions that could be approached in this fashion. In research by myself and Beryl Forgay on attitudes and beliefs of participants in the Saskatchewan uranium inquiries, reported in "Ten Years after the Uranium Inquiries," we discovered seven main themes having to do with the provincial economy, northern economy, energy policies, ecology and technology, science and knowledge, the inquiry process and the regulatory system. Once these were broken down we had over a hundred specific issues around which controversy could occur. The scope is huge, which is one reason why an uncritical public is prone to accepting simplistic clichés.

Shoddiness of Nuclear "Educational" Resources

When we look at the resources being supplied to schools on uranium mining and nuclear power, it isn't surprising that many teachers are ill informed about the downside of nuclear energy. A past issue of *Green Teacher*, "The Nuclear Industry Goes to School," gives many examples of how the industry has penetrated the school system with industry propaganda in Saskatchewan and elsewhere.[3] The examples have multiplied in Saskatchewan since we became the front-end of the global nuclear system.

In December 1990, Saskatchewan Education (now referred to as Saskatchewan Learning) released a publication entitled *Uranium* as part of its Saskatchewan Resources Series,[4] and the SMA and Uranium Saskatchewan are thanked for their "technical contributions." It was the third in a series done by the Department on Saskatchewan's natural resources — the others were on forestry, petroleum, agriculture, potash, etc. The document lists Uranium Saskatchewan and the CNA as sources of "information about Saskatchewan's uranium industry and nuclear energy," hence promoting the industry as a source of reliable educational resources on itself. Environmental and non-nuclear groups are listed only as sources of "information opposing uranium mining and nuclear energy."

Energy and Mines also issued a resource kit for teachers, with the cover printed by the government department but most of the contents supplied from the industry. The inserts include the following: "Mining... Good for Saskatchewan... Mining — What Does it Mean to Saskatchewan"; "Mining — Saskatchewan's Second Largest Industry"; and "Mining and the Environment — a Responsible Approach." All of these were provided by the SMA. Then there is an Education Series on minerals, put out by Saskatchewan Energy and Mines in 1991, and one entitled *Uranium in Saskatchewan*, produced in 1992, which has a blatant industry bias and again lists the CNA and AECL among the "other sources of information."

Let's take a closer look at the credibility of these industry-provided resources. A denial of historical fact is not good educational practice, yet industry disinformation about Saskatchewan's weapons connection permeates these resources. Until the late 1970s, there was a widespread misconception that Saskatchewan uranium had nothing to do with nuclear arms, but since then it has become known that all uranium from Saskatchewan during the 1950s and 1960s went into nuclear weapons. One estimate is that "all Saskatchewan uranium extracted between 1953 and 1969 went to help produce 27,000 nuclear weapons in the U.S.A."[5]

Yet, an SMA resource called *Mining and the Environment: A Responsible Choice* (undated) used in the schools since the early 1990s simply says, as though it is self-evident, "Uranium produced in Saskatchewan is used only to fuel nuclear power plants around the world." The Saskatchewan Energy and

Mines pamphlet, *Uranium in Saskatchewan*, in its 1991 Education Series, put it slightly differently, stating "the primary use for uranium is fuel for nuclear power reactors." But when it lists the other uses (medical, airplanes and irradiation), it fails to mention that this heavy metal is indeed used in nuclear weapons, including DU weapons. It's as if it's a taboo subject, like sexual abuse in a family.

A Teachers' Guide entitled *Uranium in Saskatchewan* was produced in 1992 by Uranium Saskatchewan, a sub-group of the SMA, which is identified as representing "nine companies which are presently exploring for uranium, developing uranium mines or operating uranium mining facilities in Saskatchewan." It was produced for grade 8 science teachers to use in the curriculum on the geology of Saskatchewan and "man's" utilization of earth's resources, and for grade 9 science teachers to use in their curriculum on energy, and energy and civilization. It is also geared for use before and after students tour Uranium Saskatchewan's mobile promotional trailer.

It is stunning to imagine this capacity of uranium corporations to direct and animate both the science teachers' and students' learning about nuclear energy. But the penetration of the industry went even further, with the AECL becoming "corporate partners" to schools in Saskatoon and Regina. This influence is analogous to the way authoritarian government's feed "political education" or theocracies feed "religious education" to students in their schools. That corporations are doing this, with government approval, is no less consequential for the independence of public education or the maintenance of a democratic civil society.

A pamphlet entitled *Uranium: The Energy Metal*, inserted in the front of the SMA's Uranium in Saskatchewan binder, for teacher's use, comes from the pro-industry Uranium Institute, based in London, England. It purports to provide a history of uranium from 1789–1989. However in discussing "the age of nuclear power," it neglects to mention the military origins of the nuclear industry. To accomplish this "propaganda device," it totally distorts and banalizes history. While it mentions the Manhattan Project, which produced the first A-bomb for use on the Japanese, it says, "After the war ended attention quickly turned to developing the peaceful uses of nuclear energy, particularly as a source of electrical energy." It then completely jumps over the nuclear arms races and discusses the expansion of uranium as though it developed to meet the needs of commercial nuclear power. It simply writes that, after the early 1950s,

> to meet the requirements of the fast-growing nuclear industry, uranium mining was expanded in the U.S., Canada, France, Australia and central and southern Africa.... These same countries, led by Canada, are still the world's major producers of uranium.

The phrase "fast-growing nuclear industry" is clearly meant to mean "the peaceful atom"; however, there was hardly any demand for uranium for nuclear reactors until the 1970s. No mention that uranium mines everywhere, including at Beaverlodge, Saskatchewan, after 1953, mined uranium solely for weapons. This historical forgery is somewhat like the Stalinists in the U.S.S.R. taking Leon Trotsky's face out of photos, so they could better propagandize the public with their revisionist history of the Soviet revolution. But Uranium Saskatchewan goes one step further and removes whole decades from its "educational" resources.

It is worth considering what the public would think of the chemical industry as a main educational source of material on "chemical facts" or the forestry industry on "forest facts." Yet Saskatchewan Education credits the CNA as a credible source of educational material on "nuclear facts." One of its pamphlets published in 1989, in the series Nuclear Facts, entitled *How Do We Protect the Environment in Uranium Mining*, shows how the industry tries to shape public opinion and school curriculum by omitting crucial information. Though the CNA admits that waste results from uranium mining, *nowhere* in the pamphlet is there any mention that the waste tailings are radioactive. Nowhere is there any mention that over 80% of the radioactivity held in the uranium ore is left in the tailings near the mine site. Nor is there any mention of the radioactive half-life of some elements in these tailings — for example, that after 76,000 years, one-half of the radioactive thorium will still be in existence in the tailings and will continue to generate highly toxic radium and radon gas.

Rather than mentioning the radioactive quality of these tailings, the CNA talks only about quantity, stressing that "uranium mines account for only 2% of all mine tailings in Canada." It also stresses that "lower volumes of tailings" come from the higher-grade uranium ore mined in Saskatchewan without mentioning that these tailings are the most radioactive of any in the world. Furthermore, there's no mention that at the time the amount of uranium tailings in Canada was approaching 200 million tonnes, which some claim is enough to cover the Trans Canada Highway, ocean to ocean, to a depth of 2 metres. From an ecological point of view the magnitude of these tailings can be catastrophic.

The CNA's depiction of these wastes is less than truthful. Its pamphlet says: "In the same way that there is a common approach to the management of used nuclear fuels, there is a common approach to the management of tailings…." This is a play on words intended to twist meanings. The only common approach for addressing nuclear wastes at present is to store them at the reactors, which amounts to "waste management" without any long-term plan. This was the case in 1989 when the CNA issued this pamphlet and it is still the case in 2007. This *ad hoc* approach is similar to that taken with uranium tailings, which, no matter what kind of engineering "state of the art" spin you wish to put on it, is really a version of dumping.

The intent of CNA's "educational" pamphlets is to give the public a sense of security that the industry has a reliable proven way to manage uranium and nuclear wastes; yet elsewhere in their pamphlet they admit that "the method used varies from mine to mine." Furthermore, the industry constantly changes its strategy about what to do with its dangerous wastes, and since 1989 the AECL has even promoted Saskatchewan taking back nuclear reactor wastes as part of a "rental" deal for the uranium fuel. Cameco now advocates this as well.

The naive perspective presented on tailings management by the CNA serves their interest. The CNA downplays the radioactive issue by saying the solid tailings are "similar in composition to the ore that was originally extracted…." It falsely asserts that, "the solid tailings are no more hazardous than was the original rock"; yet elsewhere it admits these tailings, which it fails to identify as radioactive, "are, however, more mobile" than the elements would be in the hard rock. It then says, "tailings management efforts are directed primarily to ensuring that the solid materials are contained." Apparently those who approved this 1989 "educational" pamphlet weren't concerned about these internal contradictions.

What isn't included is as revealing as what is. There's no mention of the problems the Saskatchewan industry is continuing to have with tailings management. The pumping of tailings into a mined-out open pit under the water table of Wollaston Lake constitutes an ecological experiment, as does the use of bentonite clay as a lining for tailings at Key Lake. There's no mention of the Cluff Lake mine having had serious problems regarding the "disposal" of high-level radioactive wastes, nor that these were returned to the tailings piles — a practice initially ruled out in the public inquiry.

There are similar omissions of relevant information regarding the CNA's discussion of liquid wastes, which it tries to minimize by simply asserting there will be "minimal" impacts on the environment. There's no mention at all of the fact that from 1981 to the issuing of this pamphlet in 1989, there were 150 reported spills at the three operating mines, two-thirds of them involving radioactive waste.[6] And the CNA did not remind the public of the radioactive spill at Key Lake soon after this "state of the art" mine opened in 1984 or of another large spill at the Rabbit Lake mine near Wollaston Lake in 1989. (These spills are discussed in chapters 4 and 7.)

Decoupling Nuclear Energy from Science Education

It isn't too surprising that the nuclear industry slants and omits such information. What is surprising and disconcerting is that the public education system promotes such industry resources as educational material for the teacher and classroom. The erroneous idea that the scientific community embraces nuclear power is perpetrated by such pro-industry resources being promoted in the public education system and the science teachers who use them. While it is true that scientists who work for the nuclear industry publicly support their employers, this support is not general among scientists, including a growing number of younger nuclear scientists.

In November 1992, over 1600 scientists, including a majority of living recipients of the Nobel Prize for science, issued a global warning to the 160 heads of state that was republished in the Winter 1992–93 issue of *Nucleus*. Their statement talks about ecological destruction of the atmosphere, water resources, oceans, soil, forests, living species: about the growing human population and the absolute poverty that many humans now face. Their warning is that, "Human beings and the natural world are on a collision course."

And what do they propose? They write:

> We must... move away from fossil fuels to more benign, inexhaustible energy sources to cut greenhouse-gas emissions and the pollution of our air and water. Priority must be given to the development of energy resources matched to Third World needs — small scale and relatively easy to implement... [advocating] high priority to efficient use of energies, water and other materials, including expansion of conservation and recycling.... The developed nations are the largest polluters in the world today. They must greatly reduce their over-consumption, if we are to reduce pressure on resources and the global environment.... we must accept limits to... growth.

And there is no mention, whatsoever, of nuclear energy as a panacea or as any part of a solution to the energy and the ecological crisis.

The nuclear industry and its supporters in government, business, universities and the media want the public to believe that the controversial aspects of nuclear power — whether about tailings, waste or proliferation — were invented by anti-nuclear zealots. But the United Nations Brundtland Report, discussed in chapter 7, and scientists not attached to the industry have all raised these issues many times over. Saskatchewan people pride themselves on having a progressive political heritage that helped bring Medicare to Canada. But we are becoming better known for our place at the front end of the global nuclear system. And if we are so progressive and well informed, why has the misrepresentation and misinformation of the nuclear industry been able to so easily penetrate our public education system? Why are we pulling this curtain of ignorance around ourselves and our place within the global nuclear system? Why do we strive to maintain this nuclear amnesia?

It is perhaps encouraging that Saskatchewan Learning is now involved in developing pan-Canadian resources in the life sciences on topics such as "sustainability of eco-systems." After speaking to a science curriculum consultant in 2006, it seems there's no longer the stand-alone emphasis on natural resources, leaving more of a vacuum than an imbalance in this area.[7] So, perhaps the front-court press on the Saskatchewan school system by the CNA and AECL became tempered after the failure to get a Candu-3 reactor in the province. Some staff who were very involved with nuclear industry promotions in Saskatchewan in the early 1990s ended up working with Saskatchewan utilities: for example Larry Christie went to SaskPower and Leslie Gosselin went to SaskEnergy.

Since the mid-1990s, Cameco has taken over the lead role in nuclear promotions in Saskatchewan's schools, for example, by offering scholarships to northern students entering university or technical training. But it has adopted more penetrating and "cool" methods for raising its profile in the education system, for instance, by sponsoring the Saskatoon Blues Society "Blues in the School" musician performance series. It also established the high-profile Cameco Multiple Sclerosis Neuroscience Research Centre in partnership with the University of Saskatchewan, the Saskatoon Health Region and City Hospitals Foundation, and the MS Society. (Saskatchewan has one of the highest levels of MS in Canada.) It sponsors the Saskatoon Victoria Park Summer Festival, which holds the popular annual dragon boat races. It works hard to stay connected with Aboriginal groups and has received an award from the Canadian Council for Aboriginal Business. Cameco has also sponsored the multi-cultural, highly attended "Dare to Dance" event and was a sponsor of the Aboriginal Showcase at the 2007 Juno Awards in Saskatoon.

Its public relations material clearly indicates that it considers these ongoing school and community self-promotions to be effective in keeping general support for uranium mining in Saskatchewan relatively high. One could say that in the short-term, with relatively little non-nuclear activism in Saskatchewan in the last decade, Cameco has succeeded in making uranium into a non-issue. With a uranium refinery and nuclear waste dump again being proposed for the province, Saskatchewan educators, now living at the front-end of the global nuclear system, clearly must step up and challenge the neo-colonialism that lies beneath Cameco's Teflon-image and help rebalance the nuclear debate in our classrooms, in our schools and in our communities.

Notes

1. See Jim Harding and Beryl Forgay, "Ten Years after the Uranium Hearings," In The Public Interest Series, Prairie Justice Research, University of Regina, 1992.
2. Pat Clarke, "Teaching Controversial Issues," *Green Teacher, Education for Planet Earth*, Issue 31, December 1992–January 1993, pp. 9–12.

3. "The Nuclear Industry Goes to School," *Green Teacher*, June–September 1991, Issue 24, pp. 9–15. Also see Robert Regnier, "Nuclear Advocacy and Adult Education: A Case for Counter-Hegemonic Struggle," *The Canadian Journal of Adult Education*, Nov. 1991, Vol. V, No. 2, pp. 47–61.
4. Saskatchewan Resources Series, *Uranium*, Saskatchewan Education, October 1990. This kit had not been replaced or updated as of late 2006.
5. Robert Regnier, "Saskatchewan's Nuclear Agenda," *Green Teacher*, June–September 1991, Issue 24, p. 15.
6. These occurred at the Rabbit, Cluff and Key Lake mines, according to provincial records or reported in the Regina *Leader Post*, Jan. 24, 1984.
7. Dave Elliot, personal communication, Oct. 2006.

chapter 16

DISECONOMICS OF NUCLEAR ENERGY

Lesson from Ontario's Candu

While the business lobby continued to pressure the Romanow NDP government to relocate the AECL's Candu-3 industry from Ontario to Saskatchewan, a litany of problems was facing Ontario's aging Candu reactors. Ironically, another NDP government, the new Rae government in Ontario, was trying to lift the veil on decades of "nuclear secrecy." No one envied the task of the short-lived Rae government, which inherited decades of mindless Tory support for the expansion of nuclear power. Ontario's Candu problems received coverage in the *Globe and Mail* but Saskatchewan's corporate media mostly ignored the stories. That such pertinent information stopped at our borders made Ontario seem like another country and Saskatchewan feel like it was being run by a junta that feared information leaking in from the outside world. The ramped-up AECL campaign talked of a Candu boom as though it was going to be a second coming for the province, and with the curtain of ignorance drawn around our borders, it went mostly unchallenged.

Assuming that Saskatchewan teachers would be interested in raising the curtain of ignorance, when I spoke at the Prince Albert Area Teachers' Convention in February 1992, I brought to their attention a study done in 1989 by Charles Komanoff, on behalf of *Energy Probe*, to determine the role the "aging" of Candu reactors was playing in their declining performance.[1] After excluding other influences, such as strikes and retubing of reactors, Komanoff found a 0.78% decline in capability per year for older reactors and a 0.66% decline for newer reactors. Projecting this annual decline over the estimated forty years of these Candus, the average capacity would be 67%, much lower than the official estimate of 80%.

This declining Candu performance was greatly responsible for the reduction of electrical generation in Canada in the 1990s. According to Statistics Canada figures, electrical supply from Candus fell 17% during the first half of 1990 — to 4,864 gigawatt-hours (GWh) during January to June 1990, from 5,295 GWh during this same period in 1989.[2] Ontario had a 12.5% decline over this period, the largest drop in electrical generation across Canada. The reason was, "Reduced output from Ontario's Bruce and Pickering Nuclear stations, resulting from both planned and unplanned outages, contributed most to the drop, as nuclear generated electricity fell to 27.34 terawatt-hours from 33.65 terawatt-hours."[3]

According to a timely analysis entitled "Hardening of the nuclear arteries," by Tom Adams in the November 13, 1990, *Globe and Mail*, Ontario Hydro's eight oldest reactors (at Pickering and Bruce) were operating at only 42% capacity, which was nowhere near the 80% official forecast. The need for massive repairs contributed to this reduced capacity and to increasing costs. After pressure tubes failed in 1983, it took four to five years to overhaul two reactors at the Pickering plant, at a cost of $1.3 billion. Ontario Hydro estimated it would need $1 billion more to replace pressure tubes at two reactors at the Bruce plant, with backup power during the shutdown costing $800 million.

Costly safety problems were also occurring at the Bruce Candu plant. In early 1990, the plant leaked 20,000 litres of heavy water, which was referred to in the January 25, 1990, *London Free Press* story, "Public will likely pay the piper for nuclear plant's breakdowns," as "one of worst nuclear accidents in Ontario history." This accident forced the fourth shutdown of a Bruce reactor within a few months. These shutdowns contributed to further insecurity of electrical supply. Two Bruce reactors operated at 35–55% capacity, and the average performance at Bruce was about 50%.

The AECL selectively emphasized newer more efficient plants when trying to convince Saskatchewan people of the Candu's "inherently good" performance. However, the evidence suggests that the range of performance is considerable and declines significantly with age. The AECL's sales reps in Saskatchewan emphasized the world-class performance rating of the Candu at Point LePreau, New Brunswick. After only eight years of operation, the Point LePreau plant was officially projected to have a 93% lifetime level of performance — however, this failed to consider the important aging factor. The AECL also claimed a 94.4% performance in 1990 at the Embalse Candu plant in Argentina and boasted an 83.7% performance at the Wolsung Candu plant in South Korea, claiming this inherently high performance was why they were able to make a second sale to South Korea after having no exports for a decade.

But closer to home things weren't nearly so grand. According to Thomas Claridge, writing in the January 7, 1991, *Globe and Mail*, 1990 was "A gloomy year for nuclear plants" in Ontario. He reported that the Bruce A plant, comprised of four 750 megawatt (MW) reactors, operated at 93% in 1984 but only 47% in 1990. The Bruce B plant, comprised of four 850 MW Candus, did better that year, at 80.7%, but Unit 4 at Pickering had to be shut down for part of 1990 and performed at only 40%. Pickering B operated better overall at 79%, but again, nowhere near the projected Point LePreau level.

Ontario's Candus not only faced declining capacity and costly safety issues, Ontario's taxpayers faced rate increases to cover these costs. In the fall of 1990, a full year before the Saskatchewan Divine Tory government signed its MOU with the AECL, Ontario Hydro admitted to the Ontario Energy Board that it would be raising its annual power rates by 9% plus inflation, amounting to a 37% increase over three years. When Marc Eliesen, NDP-appointed Ontario Hydro chair, finally announced a 13% rate increase (compounded, this would be a 44% increase over three years), he admitted that this was needed to offset reduced performance. As *Globe and Mail Report on Business* writer Terence Corcoron said,

> One does not need to be an anti-nuclear activist to become alarmed over Hydro's strategic and financial performance over the past decade. The Darlington nuclear station cost escalation, from 1984 estimates of $2.5 billion to the current and still-rising actual cost of $13.4 billion, has created a financial albatross. The efficiency, performance levels and longevity of the utility's nuclear plants are also turning out to be far below advertised expectations.[4]

While the Candu sales staff in Saskatchewan talked of how great nuclear power was in Ontario and the Saskatchewan media "blacked out" contrary information, Ontario Hydro was engaging in desperate business practices. The utility dipped into its contingency fund, removing $245 million in 1990, $255 million in 1991 and an estimated $289 million in 1992. The decline in the value of Ontario Hydro over this period (1990–1992) was about $1 billion.

It was in this climate of Candu unreliability, safety problems and costly repairs that Saskatchewan uranium multinational Cameco decided to buy into Bruce Power, which, since 2001, has operated the Bruce Candu plant under an eighteen-year lease from the Ontario Power Generation. Bruce Power is composed of Cameco (31.6%), Trans Canada Pipelines (31.6%), the BPC Generation Infrastructure Trust (31.6%) and two unions (5.2%) — the Power Worker's Union and Society of Energy Professionals — operating at the plant.

Clearly the price was right, and Cameco was guaranteed high profitable electrical rates within Ontario Hydro's near monopoly grid. However, the cost carried by the ratepayer for refurbishing the two reactors at Bruce by 2009 will be massive: $2.75 billion according to a report in the *Globe and Mail*.[5] It was a win/win for Cameco, as this purchase also helped stabilize demand and price for its uranium, which would have dropped had no one wanted to make the costly repairs to the fledgling Bruce plant. Uranium and nuclear corporations were becoming more interlocked, providing helping hands to each other.

A story headed "Planned hydro rates jolt Ontario business," in the September 12, 1991, *Financial Post*, reported Ontario Hydro's new chair admitting that one-half of the electrical rate increase coming in 1992 was due to Darlington. The 1200-acre Darlington project, which was to have guaranteed cheap electricity for Ontario businesses and residents, had become a "sinkhole" for the Ontario taxpayer. The costs of Darlington ended up being 380% greater than the original estimate. If the Peterson Liberals elected in 1985 had canceled Darlington, $10 billion might have been saved.[6] Had this been invested in energy conservation and efficiency, projected electrical demand would have been substantially reduced, perhaps by as much as 30%. When will we finally learn this important lesson? As it turned out, even after all this cost and debt, Darlington didn't perform well and was plagued by cracked generation shafts, which puzzled plant operators. U.S. investment house J.P. Morgan expressed concern about Ontario Hydro's practice of dipping into its contingency fund, as well as about the utility's huge debt, stating reactor unreliability as a major cause for its financial concern.

The fiscal downside of nuclear power was the AECL's best kept secret as it promoted the Candu in Saskatchewan. The AECL highlighted countries having a much greater percentage of their electricity generated by nuclear reactors. At the time France had the highest percentage of electricity coming from nuclear reactors (70%) anywhere in the world, whereas Canada only had 12%. France's integrated, state-owned corporations are involved in uranium mining (Cogema, active in northern Saskatchewan), reactor construction (Framatome) and nuclear weapons testing and waste management (CEA). This centrally planned system, renamed Areva in 2001, is touted as the world's "only company to cover all industrial activities in nuclear power." It remains the envy of nuclear technocrats the world over, including those in the AECL.

What the AECL and CNA didn't mention, but was reported in the story, "Chained to reactors," in the February 2, 1991, *Economist*, is that the state-owned Electricite de France (EDF), which runs the fifty-four reactors constructed over twenty-seven years, had a debt of 230 billion French francs by the 1990s, and since the EDF accelerated its reactor construction program after 1974, it had losses of 28 billion French francs. One EDF internal report warned of the "threats" posed by several non-nuclear energy sources, and the EDF had already converted one reactor to natural gas because it was more cost-effective.

The centralization and standardization of reactor construction in France lowered some costs, though not enough to avoid this huge debt. This standardization, however, makes the French public, who get most of their electricity from these reactors, more vulnerable. If a serious safety and/or performance problem develops in a single plant, which is increasingly likely with aging, then all reactors will have to go through a shutdown for repairs.

Realizing these problems of costly aging reactors, insecurity of supply, debt load and rising consumer rates with the Ontario Candu, I asked the teachers at the Prince Albert Teacher's Convention to consider why Saskatchewan would want to get into this deteriorating business? The Romanow NDP government giving consideration to keeping the SaskPower-AECL agreement to build a Candu-3 made no sense in terms of either energy or fiscal policy, especially because the province was already facing a massive $5 billion debt and a reduced credit rating. Luckily for the Saskatchewan taxpayer and future generations, the Romanow NDP had to listen to reason.

Ontario, however, continues to have trouble breaking its addiction to nuclear power to this day. The present McGinty Liberal government stands ready to repeat the error the Peterson Liberals made in the 1980s. In 2006 McGinty announced plans for two more reactors at the Darlington site at an arguably under-estimated cost of $3 billion. The province's full energy package, which could cost $70 billion by 2025 and thankfully includes some plans for conservation and wind power, still has to go to the Ontario Energy Review in 2007.[7]

As Ontario once again prepares to publicly debate its dependency on nuclear power, non-nuclear groups will be amassing updated information on the Candu's declining capacity, safety problems and rising costs. But this time they will come face to face with the Saskatchewan-based Cameco Corporation, the main operator of Ontario's Bruce Candu plant. The Ontario public and non-nuclear groups need to ask for a full costing of nuclear power that includes uranium tailings at the mines owned by Cameco in Saskatchewan, as well as nuclear wastes and reactor decommissioning. As the uranium and nuclear corporations become more integrated, it will become more difficult for them to compartmentalize the costs and risks of the full nuclear fuel system and to keep people in one province or country from knowing what's happening in the other.

Inflating the Demand and Hiding the Costs

From a global education and activism perspective it is necessary to look beyond Ontario and Canada. The scope of issues that could be addressed is huge, but I will focus in this chapter on economic issues, because the proponents appeal to economic benefits as their main line of argument, and we have already seen how "economic development," rather than energy resources *per se*, has been embraced as the industry's main propaganda strategy in Saskatchewan. From a balanced, multidisciplinary perspective there is the need to not only look at economic costs, which includes capital costs, decommissioning costs and costs of research and development (R&D), but also at the actual interconnections between the economy and energy demand.

It is important to carefully scrutinize sources of information for making any argument in this controversy. The sources of my information on energy costs and demand are not from the opponents of the nuclear industry but primarily the pro-nuclear Nuclear Energy Agency (NEA) of the OECD countries.

It's revealing to look back at what the NEA has said about nuclear demand in the past, and to compare this to what has actually transpired. Such an analysis in *The Economist* in 1991 concluded that, "after three decades of commercial development, nuclear power has failed to fulfill its promise." It noted that at that time: "Nuclear power provides only 17% of the world's electricity — 5% of its total energy use — from about 420 plants, four-fifths of them in rich countries."[8] The analysis also noted that in 1972, just prior to the energy (i.e., oil price) crisis, the NEA predicted there would be 1,000 gigawatts (GW) of nuclear-generated electricity in the world by 1990, but there had been only 260 GW, or 26% of that predicted. In 1990 the NEA lowered its forecast to only 300 GW nuclear-generated electricity worldwide by the year 2000. A similarly revealing failed prediction occurred with nuclear power in the U.S., where in 1972, the U.S. AEC predicted there would be a thousand plants by the year 2000. The actual number, as Helen Caldicott points out in her book, *Nuclear Power Is Not the Answer*, was only 10% of that, or 103.

Since 2000, the NEA, like the CNA, has tried to retool its sales pitch to make nuclear look like it's a "sustainable" energy source,[9] talking rhetorically and confusedly about "Nuclear Energy and Sustainable Development" on its website. Meanwhile, obviously aware of its growing cost-uncompetitiveness, it looks at ways to reduce costs in design, construction, management and by streamlining regulatory processes. But so far the growth curve in worldwide nuclear power remains fairly flat. The NEA reports that nuclear power still only produces about one-fifth of the electricity across the OECD countries, and its 2006 report indicates that plant closures and start-ups balance each other out in the foreseeable future, with nine more plants, five in the U.K., to be phased out and only ten presently under construction worldwide, four in Korea, three in Japan, two in Slovak and one in Finland.[10]

The Economist noted back in 1991 that the "rich countries have repeatedly found they needed less electricity than they had forecast," which is something we saw when SaskPower was in league with the AECL. By 1990 every country with nuclear reactors in Europe, other than France, had a moratorium on new plants. In the U.S., with nearly one-quarter of the plants in the world, there were "no new plants without subsequently being cancelled, since 1974," and it wasn't until 2002 that the U.S. announced another nuclear plant, which isn't expected to be completed until 2010.[11] So in spite of all the disinformation and hype about nuclear being "clean," there is presently no boom occurring in planned nuclear reactor construction.

By 1990, most OECD countries were already planning to get less electricity from nuclear plants by 2005. Only in Japan and South Korea, wrote *The Economist* in 1991, "does the future of nuclear power still seem promising." And Japan remains the major exception to the global trend. It started to use nuclear power in 1966, and because of its heavy dependence on imported energy, embraced nuclear energy in a big way in 1973 after the oil crisis. It now has fifty-five reactors, which produce 30% of the country's electricity, and has three new reactors under construction, wanting to reach 41% of electricity by 2014.[12] The Japanese non-nuclear movement — like the French — faces a huge challenge in trying to redirect their country to an ecologically sustainable path. Accumulating nuclear waste, the danger of nuclear accidents and recent knowledge — referred to by past U.S. president Bill Clinton when he spoke to NGOs at the Global Conference on Global Warming in Montreal in 2005 — that geo-thermal energy could replace one-half of the nuclear energy in Japan, are bound to sink in. Radiation leakage after reactor damage and a fire resulting from an earthquake in the summer of 2007 showed how ill-conceived was Japan's hasty expansion of nuclear power in the 1970s. Tepco, the operator of the damaged nuclear plant, has interests in Saskatchewan's Cigar Lake uranium mine.

Like France, Canada planned for a slightly larger percentage of electricity from nuclear in 2005 than in 1991, but changes to Ontario Hydro — with its own deceptive history of inflating demand — required Canada's projections to be reduced.[13] By 2000, according to the NEA, Canada got only 14% of its electricity from nuclear plants, which put it near the bottom of all OECD countries with nuclear power. Hydro and then coal continues to supply most of Canada's electricity.

According to the November 16, 2006, *Globe and Mail*, the expanding industrial giant China is considering thirty to forty nuclear reactors by 2020, so it may become the other exception to the global trend. However, though this sounds like a lot, this would only bring nuclear energy from 2% to 8% of China's energy supply. And in the coming years China will likely learn that energy conservation is not only cheaper but better for the environment. China is also seriously exploring wind power, which I discuss in the next chapter.

The hope that China's reactor market would be the salvation of the struggling AECL, after two Candu sales in the 1990s, collapsed when China chose the light-water design used in Europe, Japan and the U.S. It appears that the big users of nuclear power are all going with this technology. Saskatchewan is still trying to get uranium contracts with China; however, China's 2006 deal with Australia — which has eased its export restrictions — suggests China may be importing 2,500 tonnes a year from Australia in the next few years.[14]

Demand for nuclear power in the industrialized countries, where four-fifths of the nuclear plants exist, simply did not grow as the industry wanted us believe. The Candu-3 sales reps travelling to Saskatchewan's schools after 1989 told teachers and students we were lucky to be a new base for research and development in this industry; however global trends at the time suggest we were, temporarily, mesmerized by a white elephant.

And what about the costs of this industry? The nuclear industry typically emphasizes its relatively lower costs for fuel and operations compared to coal or gas-fired electrical plants. However, 50% of the cost of nuclear plants is usually for capital, which carries high interest costs and creates a major public debt load. A small shift in interest rates is enough to make a nuclear plant totally uneconomic. *The Economist* pointed out in November 1991 that while "the true cost of nuclear power is hard to calculate," the worldwide industry has long suffered from what it calls "appraisal optimism," whereby the actual costs are much larger than those originally projected.

In the U.S., actual costs of constructing nuclear plants more than doubled over those originally projected. This was true for all seventy-five nuclear reactors constructed between 1966 and 1977, when the initial capital cost estimate of $298 (1982 U.S. dollars) per KW electricity went up mid-way through construction to $414 per KW and by completion was at $623. By 1976 and 1977, the discrepancy in costs was even greater: startup — $794; mid-way — $1,065; and final — $2,132.[15] We've already seen how it's been the same story in Ontario, where at the Darlington Candu plant, costs in 1991 were $13 billion compared to the $2 billion projected in 1973.

Because of such high capital and interest costs nuclear power requires a high level of reliability of output and sales to service its huge debt. This helps explain the heavy emphasis on promoting more sales and the industry's traditional hostility to, and disinformation about, energy efficiency and conservation, or demand side management. But, as we've seen, the reliability of output of nuclear plants is seriously in question.

The relatively low cost of uranium fuel is no longer an advantage. Back in 1991, *The Economist* said, "The biggest recent change in the economics of nuclear power has been the collapse in the cost of rival fuels.... True, uranium prices too have tumbled, but they make up only a small part of the overall cost of nuclear power." The article quotes one nuclear proponent saying, "Anywhere there is natural gas, nuclear will have to mark time." And while the price of natural gas has risen sharply since 2000, the price of uranium, which recently went from $55 to $85 a pound, has risen even more steeply.

Nothing much changed after 2000. Few European countries are considering new nuclear power plants. In Britain, the first country to produce nuclear power, the government's Performance and Innovation Unit (PIU) reviewed energy policy in 2002, and, according to Rob Edwards in the *New Scientist*, "recommended that nuclear power should be retained only if expanding renewable energy sources and improving energy efficiency don't work." A leading economist with the PIU even commented "that in Britain a nuclear power plant that could compete economically with other forms of energy has never been built."[16]

The Economist pointed out in 1991 that the most unaccounted for costs of nuclear power have to do with nuclear wastes and decommissioning, noting that "the need to dispose of radioactive waste has always been a problem for the industry.... No country has yet taken a final decision on where to put its high-level waste." It also notes that, "when a plant is decommissioned, its highly radioactive core has to be disposed of.... So far, no one has squarely faced this choice, since no commercial plant has yet been decommissioned."

The statement on nuclear waste, written in 1991, remains totally applicable today. Unable to create a credible proposal for long-term nuclear waste "disposal," the industry is now trying to transform the nuclear waste problem into a problem of perception. And the diseconomics of the industry has already led to some plants closing before they reached their "old age." Already by 1991, according to *The Economist*, regulators encouraged "utilities to compare the cost of continuing to use existing plants with other ways of increasing supply or reducing demand. Compared with energy-efficiency light bulbs — indeed, often compared with coal, nuclear suffers." And the challenge of decommissioning reactors continues to be underestimated through a magical discounting of future costs. Where nuclear plants have been decommissioned in the years since 1991, costs have greatly outstripped projections. In her 2006 book, Caldicott points out that decommissioning the Yankee Power reactor in Massachusetts was projected to cost $120 million but ended up being $450 million.

With the shutdown of over 400 nuclear power plants and the challenge of stewardship of accumulated wastes worldwide, the days of reckoning steadily approach. And the lack of a "permanent" waste storage or proven decommissioning system raises hard questions about intergenerational justice. Are the risks and costs of nuclear technology being passed to our children and grandchildren and great grandchildren, while the benefits are being gleaned by only a few for the very short term? There is no sign of foresight in the way this industry has developed. Its "out-of-sight, out-of-mind" mentality clouds over these vital issues. The fact that Saskatchewan is now the largest supplier of nuclear reactor fuel in the world, with no permanent waste storage in place anywhere, will continue to raise profound moral and ecological questions for citizens of this province and country. This is something members of our present Calvert NDP government or a future Saskatchewan Party government might want to quietly and honestly contemplate.

Projected costs of nuclear waste storage and decommissioning were the main reason that the Thatcher government in the U.K. couldn't privatize its nuclear electric plants in the 1980s. About this, the 1990 Energy Committee of the British Parliament wrote: "The government was seeking to expand an uncompetitive source of power in a competitive market."[17] By 1991, the average operating and maintenance costs for nuclear plants in the U.S. was higher than for fossil-fuel plants. And though the U.K. was planning to phase out, by 1998, an 11% levy imposed on all electric bills (a hidden subsidy to nuclear energy), it was unclear how it would pay for the cost of dismantling eight "ancient Magnox plants."

The U.K. still has twenty-three reactors producing 20% of the country's electricity. However, Britain's last nuclear plant was completed in 1995, and the Labour government's energy policy review doesn't recommend more. However, the unrelenting nuclear lobby is still hard at work. And it is in more collaboration with the U.S. nuclear industry, with even closer military-industrial ties between the U.S. and U.K. administrations since the invasion of Iraq in 2003. A secret report done for the British Department of Trade and Industry calls for loosening up nuclear plant approval procedures and huge carbon tax breaks to allow a nuclear comeback, but it notes with concern that tightening the regulations on radiation releases into the seas could prevent new nuclear plants from being built. Writing in the *New Scientist* in 2002, Rob Edwards concludes that without these "reforms and the carbon tax breaks, nuclear power would flounder."

By 1991 *The Economist* reported: "Rising costs of nuclear power have already driven Mexico, Brazil and Argentina to shelve their ambitious nuclear programmes." Regarding the prospects for nuclear power in developing countries in general, it concluded: "Capital-intensive, complicated and risky nuclear power is not yet an appropriate source of energy for most developing countries." This prognosis had major implications for the Candu-3, promoted so strenuously in Saskatchewan and which the AECL hoped to sell to the Third World, and in the late 1990s this design was finally put to rest without one prototype being constructed.

It's much the same story when we look at the hidden costs of nuclear research and development. From 1980 to 1991, OECD countries spent $90 billion (in 1991 dollars) on nuclear research, and they did this, according to *The Economist*, "without getting anywhere much." It is quite revealing and startling how public spending on nuclear technology over recent decades contrasts with spending on renewable sources of energy and conservation, upon which sustainable development will depend. The bulk of spending on energy research — 59% on average from 1980 to 1991 — went to nuclear, including conventional, breeder and fusion research. The percentage for 1991 was still over half, at 54%. Only 8.9% of the spending — on average, from 1980 to 1991 — went to all renewable sources. Even less, 7%, went to renewables in 1991. And even though it has been proven repeatedly that it is much cheaper to reduce demand than increase supply, only 16.5% — on average from 1980 to 1991 — went towards conservation research. It was some sign of hope that OECD spending on conservation went to up 22% in 1991; however, non-renewables still predominated, with the rest of the research funds, 15–16%, going to fossil fuels.[18]

This is hardly a level playing field or any sort of "free market." Taxpayers throughout the OECD massively subsidize an energy source that after forty years only provided 3–5% of total energy. In 1991 the U.S. had 20% of the world's nuclear plants and produced 33% of the world's nuclear-generated electricity, so a look at its public subsidies is instructive. A 1993 study done for Greenpeace by energy economist Charles Komanoff found that commercial nuclear power had already cost the U.S. $492 billion, with $97 billion of that coming in the form of federal subsidies.[19] These were admittedly underestimates of both total and public costs, since in Komanoff's study, reactor decommissioning costs only included the $3.3 billion that had been set aside for this, with $1.4 billion set aside for reactor deficiency. Meanwhile, the official estimates for decommissioning only six reactors, out of the more than one hundred U.S. reactors in operation at the time, exceeded the $1.4 billion, and other reactor decommissioning was estimated to cost $2.6 billion. The U.S. Nuclear Information and Resource Service in its February 1, 1993, issue of *The Nuclear Monitor*, claimed that decommissioning and short-term radioactive waste storage "will itself cost $500 billion in current dollars," which they called "a financial scandal comparable only to the savings and loan crisis," which would double the cost of nuclear power without creating any more energy. This industry would clearly not have survived in the heartland of private enterprise without this level of hidden public support.

The situation in Canada, if anything, is worse. In 1979, 64% of all federal R&D grants went for conventional nuclear power, while only 21% went for new energy sources and conservation combined. The imbalance and bias increased under the Mulroney government, with federal grants to conventional nuclear energy corporations doubling from $100 to $200 million a year over his term of office. The Economic Council of Canada estimates that by the late 1980s, $19 billion (in 1990 dollars) of federal subsidies had gone to the nuclear industry in Canada, which includes research and development grants, loan guarantees and bailouts for reactors and heavy water plants. The study concludes: "It is doubtful if these costs will ever be recovered."[20] And, even with all this public support, at the time the nuclear industry still produced only 3% of the total energy and 13% of the electricity that Canadians use.

And the hidden costs don't end here. In Canada the *Nuclear Liability*

Act exempts the industry from any liability over $75 million, and we know nuclear accidents can cost much, much more. For example, by 1991, the Chernobyl clean-up had already cost $18 billion, discounting the "cost" of human suffering for the many thousands who were disabled or had already died due to diseases linked to the accident. This subsidizing of market insurance gives nuclear energy an unfair advantage over non-nuclear competitors and conservation. In 1990 Ontario Hydro and New Brunswick Power both admitted that, "striking down the Nuclear Liability Act would put at risk the continued use of nuclear energy for peaceful purposes by utilities, suppliers, universities and research institutes."[21]

In her 2006 book, *Nuclear Power Is Not the Answer*, Helen Caldicott provides more updated costs for nuclear energy. She reiterates that the omission of capital costs is the way the industry maintains the myth that nuclear power is cheaper than other sources of energy, noting that more complete costing by the *New Economics Foundation* in 2005 found "the cost of nuclear power has been underestimated by almost a factor of three." She also notes that in 2005, the *New Scientist* cost analysis, accounting for the exorbitantly high costs of constructing new nuclear plants, put the cost at 14 cents (U.S.) per kilowatt hour, rather than the 3 cent figure commonly used by the nuclear industry. These real costs continue to be hidden by huge government subsidies and handouts. Caldicott suggests: "This socialization of nuclear electricity within a capitalist society has never been called into question, nor has it been scrutinized by the general public and their elected representatives." The main reason for this socialization of costs is that "nuclear power was and still is an offshoot of the nuclear weapons industry."

The extent of the nuclear industry's dependence on government is almost beyond comprehension. Caldicott reports that 60% of all U.S. government R&D — $70 billion of the $111 billion spent between 1948–98 — went towards nuclear. In the OECD, by 1992, the figure for nuclear subsidies had reached $318 billion. Brand new subsidies totalling $13 billion were announced in George Bush's 2005 U.S. *Energy Bill*. The Bill also reauthorizes the *Price-Anderson Act*, which guarantees, as Caldicott says, that the taxpayers "will pay 98 per cent of up to $600 billion government insurance in the event of a worst case nuclear meltdown." Europe has similar accident insurance subsidies. In 2004 the Paris Convention was revised to provide public protection for nuclear liability. Caldicott comments that, "In France, if Electricite de France had to insure for the full cost of a meltdown, the price of electricity would increase by about 300%."

We have already seen that subsidies for renewable energy have never been equitable. Caldicott discusses a U.S. study in 2000, *Federal Energy Subsidies: Not All Technologies Are Created Equal*, which suggests that during its first fifteen years of development, nuclear received thirty times the financial support as wind per kilowatt hour, even though wind power provides 5 times the jobs and 2.3 times the electricity per amount of investment. State subsidies continue to keep nuclear power afloat, even though renewable energies now provide more of the total energy. According to a 2004 United Nations Environment Programme publication, "In the OECD most energy subsidies are still concentrated on the production of fossil fuels and nuclear power, although the amount of these subsidies appears to have declined in recent years."[22]

Caldicott also lists, as hidden costs, "off budget" subsidies such as preferential tax treatment and "stranded cost recovery" provided to nuclear utilities because nuclear power is unable to generate the rate of return predicted during start-up. And this doesn't even begin to account for "externalized costs" such as environmental and health damage or proliferation. Privatization and concentration of ownership should be resisted because they threaten to make nuclear costs even more invisible from public scrutiny. We are having difficulty effectively challenging nuclear propaganda about costs in part because the nuclear network has already become so ingrained in our society. As Caldicott puts it: "The intricacy, complexity, and sheer vastness of this nuclear network is enough to take one's breath away. It is so well funded, so entrenched, so organized, so out-of-sight...." This is becoming the case with the uranium/nuclear network in Saskatchewan, as shown by the pervasive influence of Cameco within mainstream institutions.

Winding Down the Canadian Nuclear Industry

The Harper minority Conservative government's attempt to make Canada an "energy superpower" through mammoth heavy oil extraction for U.S. security of supply is taking us further down the non-sustainable path, so the shift of government policy to create a level playing field for the renewable energy sector has yet to be won. This matter will likely become vital in the next Canadian federal election, where — in the face of growing concern about climate change — energy and environmental policy seem destined to be front and centre. In 2006 the Harper Conservatives announced they were considering selling the AECL, excluding its Research and Development branch with 300 scientists. The pretence, as reported by Eric Reguly and Andrew Willis in the November 16, 2006, *Globe and Mail*, was that this was a good time to do so "as the nuclear industry springs back to life." But the fact is the global nuclear market is not expanding as the industry hoped and is getting more competitive. U.S. General Electric and Japan's Hitachi previously announced they were going to pool their nuclear units in a $2.6 billion enterprise, in the hope that they "will capture more contracts."

And the AECL's books aren't impressive. Though this Crown corporation has received $21 billion in subsidies over five decades — which Tom Adams of Energy Probe says totaled $74.9 billion in terms of compound interest and real debt — in 2006 it only had assets of $1.1 billion and a profit of $5 million, out of $400 million in sales. Things don't look so good in the AECL's future, out of the running in China's market and having no guarantee it will even get the $2.5 billion contract proposed for new reactors in Ontario.

And who would the lucky prospective customers be? Apparently the French uranium corporation Areva, which has constructed a hundred reactors compared to the AECL's thirty, "pitched the buyout to the [previous] Liberal government." However Areva has a back seat to an AECL-led consortium made up of SNC-Lavalin, G.E. Canada, Babcock and Wilcox, and Hitachi, which would certainly fit in with a continental nuclear energy and military strategy. But the Harper Conservatives apparently prefer Bruce Power, which, along with TransCan Corporation, includes the Saskatchewan-based Cameco. With this possibility, no wonder Cameco officials keep pushing a uranium refinery, nuclear power and a nuclear waste dump in Saskatchewan. We must be on guard for the newest "save AECL" strategy returning to our doorsteps.

Greenpeace went after briefing notes of officials in the Ministry of Natural Resources soon after the Harper Conservative minority government took federal power in 2006. These notes, discussed in the article "Harper embraces the nuclear future," by John Geddes in the May 7, 2007, *Maclean's*, confirm that Harper considers nuclear to be "an important building block of long-term energy policy framework." Harper's Natural Resources Minister Gary Lund is quoted in Geddes' article as saying, "From purely an environmental perspective, for no other reason, you have to consider nuclear." The nuclear mandarins have clearly gotten to him, for Lund chants the totally unfounded mantra that "we're the world leader in safe management of fuel waste," while idealizing France's 80% dependency on electricity from its fifty-eight reactors, saying "they have the cleanest air shed of all the industrialized countries." Clean, apparently, includes cancer-causing radioactivity.

Geddes remarks that Harper has quickly gone from an avowed climate change skeptic to an avowed pro-nuke. My grandmother used to warn me that, "two wrongs don't make a right," and Harper's and Lund's new-found "environmentalism" is exposed by their strategy for saving the AECL. While Harper would rather let contentious energy politics be fought at the provincial level, there's no doubt his government wants a potentially privatized AECL to get a jump-start by winning Ontario's contract for two new reactors at Darlington. But this might not occur as both France's Areva (which is building Finland's new reactor) and the U.S.'s Westinghouse (which has reactor contracts in China) have advanced much further with the new generation of reactors. Design work for AECL's Advanced Candu Reactor likely won't be done until 2010, and the Ontario McGinty Liberals want the new reactors to be on-stream by 2015.

The Candu is also being promoted to the U.K.'s Labour government, which in spite of evidence to the contrary, still flirts with new nuclear reactors. However Harper's big hope for an AECL comeback is in his Alberta homeland, in the oilfields, where his Minister Lund is leading the charge to have nuclear used as an alternative to natural gas as a means to generate steam for extracting heavy oil. (I discuss this plan in more detail in the next chapter.) This is part of the deceptive strategy to make Canada, in Harper's words, an "energy superpower" — which translates into expanding Alberta's profitable and polluting heavy oil export industry.

In their convoluted approach to energy policy, Harper and Lund make no distinction between the toxic non-renewable nuclear and renewable (or soft) energy. They are all simplistically called "eco-energy." Anything that is good for business — e.g., biofuels —will be embraced, regardless of its real environmental costs. Geddes also notes that, like the nuclear industry, Harper is "looking to exploit cracks they see in the wall of environmental opposition" to nuclear. Reducing the use in the oil fields of declining natural gas reserves may get some support for nuclear from a public that is worried about shrinking gas supply for home heating while not knowing of the practical alternatives. But a little thought will go a long way: nuclear energy in the oilfields would be creating more heavy oil and huge amounts of greenhouse gases.

Saskatchewan people know first hand of such deceptive nuclear strategies, for we endured them during the AECL's onslaught to get the Candu-3 in Saskatchewan in the late 1980s and early 1990s. The deception apparently continues to move west, from Ontario to Saskatchewan to Alberta, and Saskatchewan non-nuclear groups will now have to join forces with non-nuclear Albertans committed to sustainable, renewable energy. Meanwhile, as happened back in the 1980s and 1990s, it is public subsidies that continue to keep the AECL afloat. According to Geddes, in 2006, another $180 million went to the AECL and a $520 million/five-year commitment was made to clean up wastes at AECL research sites. Responsible and intelligent energy policies in the past could have saved us, and future generations, of all this trouble and risk.

According to an Ipsos Reid poll discussed in the *Maclean's* article and done for the CNA, the deception of the nuclear industry may be somewhat successful in the short-term. The poll found that over the last two years support for nuclear power in Canada rose from 35% to 44%. In Ontario, the only place where any dependency on nuclear power exists, it went from 48% to 63%. This may be due in part to continuing fears about there being an insufficient source of electricity, especially after a large Ontario power grid failure in 2003. But the public understanding and perspective on how to create sustainable, non-polluting energy is quickly evolving. Common sense works for the renewable energies: having decentralized electrical generation at several points on the grid actually makes it less vulnerable to going down. And even with Ontario's announced addition of two new reactors, by 2025 — within its overall energy mix — Geddes reports that electricity from renewables (15,700 MW) will outstrip that from nuclear (14,000 MW). With the ongoing technological revolution in renewables, this will likely prove to be an underestimation.

The CNA's public opinion findings are not that supportive considering that the nuclear industry has had an almost "free ride" of late in the public energy debate. Even with the slight shift towards support for nuclear, probably because it is falsely being seen as "clean," it is still a minority of Canadians (44%) that supports its expansion. If non-nuclear groups across Canada are able to do fundamental education to expose the latest series of nuclear myths and show the practical alternatives, public opinion will again shift back to a non-nuclear view, as it did in the late 1970s, after the nuclear industry tried and largely failed to exploit the "energy crisis." Exposing that nuclear is not a magic bullet for climate change, but is in fact a huge user of fossil fuels will be part of this. And the non-nuclear movement must show the connection between nuclear power and nuclear weapons, which I discuss at length in the last chapter. And this time, unlike in the 1970s, the economic data is overwhelmingly against nuclear — which could be the final nail in the nuclear coffin. Saskatchewan — now the front-end of the global nuclear system — is where this battle over economics and sustainability must now be waged and won.

Notes

1. See discussion by Tom Adams, "Study Shows Candu Reactor Performance Declining with Age," *Ippso Facto,* Nov.–Dec., 1989, p. 6–7.
2. Statistics Canada, Bulletin 57-001, Supply and Services, Government of Canada, 1990.
3. "Electric Generation Down 5.4% Through First Half," *Energy Analects,* October 15, 1990, p. 5.
4. "Signs of colossal failure at Ontario Hydro," *Report on Business, Globe and Mail,* July 13, 1991, p. B2.
5. Eric Reguly and Andrew Willis, "Ottawa eyes sale of AECL," *Globe and Mail,* Nov. 16, 2006, p. A1, 11.
6. G. Scotton, "Screwed up nuclear plant symbol of utility in trouble," *Financial Post,* July 29, 1991.
7. Steve Erwin, "Nuclear power advised to meet future needs," Dec. 9, 2005, available at <www.integrityenergy.ca/content/view/55/44/> acessed July 2007.
8. "Nuclear power: Losing its charm," *The Economist,* November 2, 1991, p. 21.
9. See <http://www.nea.fr> (accessed June 2007).
10. Nuclear Energy Data, NEA website, 2006. (?can you get a specific web-site?)
11. Rob Edwards, "Secret plan to revive UK nuclear power industry," *New Scientist,*

July 3, 2002, available at <http://www.newscientist.com/article/dn2499-secret-plan-to-revive-uk-nuclear-power-industry.html> accessed June 2007.

12. "Nuclear Power in Japan," Briefing Paper #79, Apr. 2007, available at <http://www.uic.com.au/nip79.htm> accessed June 2007.
13. B. Verdun, "There will be no more Ontario Hydro mega-projects: Future electrical generation will be in small plants," *The Independent*, National Edition, February 1993, p. 5.
14. Geoffrey York, "China sets sights on Canadian uranium supply," *Globe and Mail*, pp. B1, 12, Nov. 16, 2006.
15. "Nuclear power: Losing its charm," *The Economist*, November 2, 1991.
16. Rob Edwards, "Secret plan to revive UK nuclear power industry."
17. *The Cost of Nuclear Power, Vol. l.* Report of the Energy Committee of the British Parliament, London, 1990.
18 "Nuclear Power: Losing its Charm."
19. C. Komanoff and C. Roelofs, Komanoff Energy Consultants, Greenpeace, Washington, D.C., 1993.
20. George Lermer, *AECL: The Crown Corporation as Strategist in an Entrepreneurial Global Scale Industry*, Economic Council of Canada, 1987. A more recent analysis of subsidies to AECL has been done for the Campaign for Nuclear Phase-Out. See David H. Martin and David Argue, "Nuclear Sunset: The Economic Costs of the Canadian Nuclear Industry," Campaign for Nuclear Phase-Out, Ottawa, 1996.
21. "Nuclear power is expensive," Campaign for Nuclear Phase-Out, November 1990, p. 2.
22. Anja von Moltke, Colin LcKee and Trevor Morgan (eds.), *Energy Subsidies*, UNEP, 2004.

chapter 17

OUT OF THE NUCLEAR CLOSET

Foot-dragging on Kyoto

In January 2001, Lorne Calvert — a past NDP MLA who ended up working for Roy Romanow in the Executive Council — was elected NDP leader and premier. He returned to the Legislature after winning a by-election in Saskatoon, which headquarters many uranium companies. Calvert would have known of, and perhaps even been party to, Romanow's political shenanigans over the JFPP on new uranium mines. Still, many non-nuclear activists hoped his election signalled a shift in direction on energy and environmental policy, where Saskatchewan could begin to emerge from its nuclear nightmare and contribute to ecological sustainability. We were pleased, soon after, when SaskPower brought the first wind-farm (at Gull Lake) into the Saskatchewan grid. But this proved to involve some window dressing. With the NDP government so heavily dependent on the oil and gas industry as well as uranium, the Calvert government stalled on signing the Kyoto Accord.

Since the Earth Summit in Rio de Janeiro, Brazil, in 1992, pressure has steadily grown for an international agreement to reduce greenhouse gases. Initially it looked like Canada would stick with other oil-dependent societies, most notably the U.S., in resisting such an agreement. However, near the end of his tenure, facing strong pressure from Europe, Prime Minister Jean Chrétien indicated his Liberal government would sign the Kyoto Accord. The Harper Conservatives, in opposition to the government over Kyoto, backed their continentalist neo-conservative allies in the White House; the Bloc Québécois and federal NDP were strongly in favour of the Accord, but again the Saskatchewan, now Calvert-led, NDP government was out of step.

With its foot-dragging on the Kyoto Accord, the Saskatchewan NDP government was paying the price for twenty years of procrastination over energy policy. While NDP uranium proponents had blindfully argued nuclear power was the alternative to oil dependence, in-depth analysis consistently showed otherwise. So the Saskatchewan government ended up dependent on both non-renewables — uranium and fossil fuels — with little "wiggle room" for alternative energy development.

Saskatchewan NDP governments should have been more proactive on energy policy since the scientific debate about the major role of fossil fuels in global warming has been going on since the 1970s, and the indicators of dramatic climate change (increases in global average temperature, melting of glaciers, deterioration of coral reefs, upsurge in forest fires, etc.) have been there for well over a decade for anyone who wants to see. Yet, after all these ecological forewarnings, Saskatchewan remained second only to Alberta in its dependence on the fossil fuel industry. In 2002, after an economic downturn and threat of a new provincial deficit, Saskatchewan again lowered its royalty rates to the fossil fuel industry to try to stimulate growth in this sector. Hanging a lot of its hopes on the continuation of the fossil fuel economy, the Calvert NDP government helped fund the University of Regina's Petroleum Research Institute, while SaskPower continued with its research on "clean coal."

Meanwhile, the Kyoto debate refocused attention on the untapped potential of sustainable energy, following on a growing worldwide view that in addition to conservation and efficiency, solar, wind, hydrogen, biomass, tidal and other renewable energy sources are the practical, least-cost alternative to fossil fuels. This is very different from the 1970s, during the so-called "energy crisis," when elites and technocrats stridently and naively promoted nuclear power as the alternative to oil. This controversy was most intense in Saskatchewan, where the NDP Blakeney government, in power from 1974 to 1982, tried to cash in on the hoped-for growth of nuclear power by creating a publicly owned uranium mining company, which later became the uranium/nuclear conglomerate Cameco. Taking over the huge deficit of the Devine Tories after two terms out of office (1982–1991), with the province's Heritage Fund depleted in part from subsidizing the uranium industry's infrastructure, the Romanow NDP government ultimately sold off its last share in Cameco. As a result of this shortsighted "public enterprise" policy, the Blakeney NDP's legacy to future generations will be accumulating uranium tailings in the north and reactor wastes abroad.

The calamitous energy policy held the province back from developing sustainable renewables. Much public money and valuable time has been squandered on the uranium ordeal. Not until 2001 did we see the first wind power project in our windy province; meanwhile, other places had forged ahead with all sorts of innovations in renewable energy technologies. In 2007 in Saskatchewan, in contrast to many other jurisdictions in North America, you still couldn't sell excess wind power to the provincial grid, which would encourage decentralized electricity generation and ecologically inclined rural diversification. More than a decade after its own Energy Options Panel called for conservation and efficiency, SaskPower had barely started to embrace demand-side management or tapping potential sources of wasted energy along the grid. In fall 2006 SaskPower finally announced it would buy 15 MW of electricity from heat recovery at Alliance Pipeline Limited. This is still connected to creating greenhouse gasses, but it is a start.

The same kind of thinking that supports centralized non-renewable energy has led to an emphasis on chemical farming, pig factory farms and other unsustainable practices instead of moving towards organic, sustainable agriculture. Very late in his first full term of office, in May 2006, Calvert finally acknowledged the upsurge of organic farming in Saskatchewan by appointing a legislature secretary in the area.

Even though Calvert's NDP government tried an environmental face-lift with wind energy, it remains fundamentally committed to and dependent upon the old polluting, non-renewable energies. While SaskPower asks sympathetic consumers to pay a little more for its "green" wind power, it committed more than half a billion dollars to upgrade the coal-burning plant at Coronach. And the Saskatchewan government pushes ahead with gas and oil exploration. In 2002 it allowed a record 1700 gas wells to be drilled in the province. By 2004, Saskatchewan produced 155 million barrels of oil, 18% of Canada's total, and in 2005 another record 3750 oil wells were drilled.[1]

The "future is wide open" pre-2003 election campaign highlighted the 20,000 oil wells presently operating in Saskatchewan, which from the point of view of the devastation brought on by climate change, isn't something to brag about. There was no mention in this campaign, or in ads for the province's 2005 centennial, of the expanding uranium mines in the north. Perhaps the government's ad handlers didn't want to ruin our self-congratulatory celebrations by disturbing our nuclear amnesia. Nevertheless Saskatchewan continues to expand both the radioactive nuclear fuel system and fossil fuels, with a big chance of being left behind with our globally destructive white elephants.

The Calvert NDP government had strange anti-Kyoto bedfellows, Alberta and Ontario, when it initially stated it wouldn't sign the Accord. Alberta Premier Ralph Kline must have chuckled when the principled, Medicare-defenders across his eastern border showed their capacity to also rationalize their anti-Kyoto, oil-rich opportunism. This regressive stand made Kyoto the big issue at the 2002 NDP provincial convention, where delegates overwhelmingly supported ratification of the Accord; afterwards the government continued going in the other direction.

This situation was reminiscent of 1982, just before the defeat of the Blakeney NDP by the Devine Tories, when the NDP government's commitment of public funds to uranium expansion, rather than conservation and renewables, deeply alienated the growing environmental and non-nuclear movements. I noted this similarity in "NDP and Energy: Shades of 1982?" published in the Regina *Leader Post* on December 24, 2002. Apparently learning little from the past and still out of step in 2002 with ecologically and socially responsible energy policy, the NDP government opposed Kyoto. Without speculating on whether the political outcome in the upcoming (2007) provincial election will be similar to 1982, it is safe to assume that the Calvert NDP's baulking on Kyoto and its promotion of a uranium refinery won't help keep its fragile electoral coalition intact. Certainly the Saskatchewan Party, like the Tory party before it, is not likely to embrace ecologically progressive energy policies, so the Saskatchewan electorate once again risks losing either way. And yes, it's unfortunate that we don't have proportional representation so that "green politics" could push the seemingly suicidal Saskatchewan social-democratic NDP into more insightful energy policies, as has happened in Europe, where a nuclear phase-out continues in several countries.

The biggest irony in all of this is that it was the Chrétien Liberals, not notorious for environmental leadership, that created the incentives for the first wind farm in Saskatchewan, and it was they who challenged the Calvert NDP to get it right on Kyoto. If newly elected federal Liberal leader Stéphane Dion becomes prime minister after the next federal election, we may see a greened Liberal administration in Ottawa.

Centennial Action Plan on the Economy

By the time of Saskatchewan's 2005 centennial, the Calvert's government's vision had even greater dependence on the export of oil and uranium. The much-touted *Saskatchewan Action Plan for the Economy*, released in September 2005, talked sparsely of "a green and sustainable economy" and of "21st century energy sources," meaning "hydrogen, ethanol, biodiesel and other biofuels — to further supplement our energy base...." The market-driven production of ethanol, where you can get more money for corn and grain going into fuel for machines than food for humans, does not really constitute a "sustainable economy."

This "pie in the sky" and rhetorical "green" visioning was quickly overshadowed by the realities and consequences of NDP energy policies, for the *Action Plan* noted that 78% of Saskatchewan's primary energy production in 2003 was for uranium. That the province was this deeply into facilitating nuclear power even astonished the non-nuclear movement. The next highest energy source was fossil fuels, at 22% (comprised of crude oil at 14.3%, natural gas at 4.9% and coal at 2.4%). The production of renewable energy only made it into the statistics because wind power was combined with hydro and wood — even so, only comprising .4% of Saskatchewan's primary energy production in 2003.

After more than half a century of CCF-NDP governments, Saskatchewan was at or near the bottom of Canadian provinces for creating renewable, sustainable energy. Choices made and directions taken by Blakeney, continued by Romanow and embraced by Calvert have left Saskatchewan as an outpost of non-renewable energy production, which will most certainly threaten future generations here and abroad.

The *Action Plan* was an opportunity for the Calvert NDP to prepare the Saskatchewan people for the challenges of energy conservation and renewables. Rather than doing this, it engaged in semantic (public relations) manipulations of the government's energy policy contradictions. Rather than talking of the need for renewable energy as a strategy for ecologically sustainable development, it says: "Our vision is to link green technologies such as wind and solar power with economic growth, bringing more jobs and opportunities to Saskatchewan people while protecting the environment." But there's no follow up on the high potential for such jobs from the renewable sector. Rather it shifts perspective and engages in high-tech rhetoric about the mining industry being "a major user of advanced technology such as virtual technology and three-dimensional seismic information to explore for new deposits," as though this has anything at all to do with sustainable development. The vision of the *Action Plan* regarding energy is business as usual, for it emphasizes expanding "uranium, potash and other mineral production in response to rapidly growing global demand...."

The *Action Plan* has a lengthy section on uranium, stressing that "Saskatchewan is the only jurisdiction in Canada that produces uranium — supplying 30 percent of the world's output." The plan commits Saskatchewan to

> stimulate further uranium exploration and development... evaluate opportunities for processing and refining uranium in Saskatchewan... [and] as a major producer of uranium... consider adding value to this resource....

This is not justified as helping develop a sustainable energy policy but solely as lucrative economic development. Yet the jobs that come in the uranium industry come at a huge ecological expense and frustrate our search for a peaceful future. Neither are they a way to "build a sustainable economy," as each of them costs (minimally) $750,000 in capital investment. Where is the sustainable job creation strategy in that?

Selling Oil and Uranium to the U.S.

The Calvert NDP's shortsightedness has been catching up to it. The contradiction of saying it supports nuclear as an alternative to oil while at the same time expanding oil has, in particular, come to haunt it. Perhaps that is why Premier Calvert said it was "all business" after he travelled to Washington in February 2006 to sell his non-renewables to U.S. Vice President Dick Cheney. Calvert and his Industry Minister Eric Cline wouldn't be raising any fuss about the Pentagon contaminating Iraq with Saskatchewan-originating radioactive DU or in anti-tank bombs, or mentioning that some international law specialists have targeted Cheney as a candidate for charges of war crimes as one of the instigators of the illegal invasion and occupation of Iraq.[2] No, this was a business trip to sell Saskatchewan's energy resources to the largest market on the planet, with no questions asked.

Calvert came fully out of the nuclear closet earlier in the month when he gave his State of the Province address to the Greater Saskatoon Chamber of Commerce. It was near the fifth anniversary of his swearing in as premier, and he had seemingly made the full transition from prairie social democrat to branch-plant resource comprador. As a *Star Phoenix* writer said after the speech, "Calvert has crossed his particular Rubicon… especially the idea of Saskatchewan's nuclear future. Now he wants nuclear companies to consider refining uranium here…."[3] The Saskatoon reporter commented on Calvert being ready to go "forth to America to be a booster for Saskatchewan's energy wealth," saying "you wouldn't know the Saskatchewan Premier was from that side [the Canadian left] of the political spectrum the way he gushed on" about how our resources can be American resources. "As his new friend Lorne Calvert is about to tell him, Cheney and the United States can keep the lights burning and the SUVs running thanks to their new friends in Saskatchewan."

The NDP grassroots, who had hoped Calvert's windmills were more than political tilting, remain puzzled by the adamancy of his nuclear rhetoric. How could this have happened to a man from the United Church ministry, which stresses stewardship and ecology, a man who as opposition Environment Critic mingled with 200 delegates from twenty countries working for a sustainable non-nuclear future at the International Uranium Congress, in the same city some nineteen years ago? It's not as simple as "power corrupts"; however, tenuous power in a continentalist market can definitely be blinding.

Reasonable rationalizations for politicians doing big business can always be found, and Calvert's dumbed-down reasons clearly come from the nuclear promotions that claim nuclear energy is the way to stop global warming. In justifying a uranium refinery, he told the Saskatoon Chamber of Commerce: "Now is the time to begin thinking about seeking that refinery capacity…. I think these are new times. There is more concern now about global warming." Of course, that's not what he said when he baulked on endorsing the Kyoto Accord a few years before. Nor was that even implied when he appointed MLA Peter Prebble to study the potential of renewable energy for Saskatchewan. Calvert, of course, also had some social democratic populist rhetoric to throw in, talking of leaving no one behind in the economy of corporate-government collusion.

The gist of Calvert's appeal to U.S. business was that Saskatchewan is "the province with the most diverse energy resource base in Canada," while emphasizing that the province is Canada's "second largest energy provider for oil and will continue to be a reliable supplier of uranium."[4] Calvert touted Saskatchewan's estimated nineteen billion barrels of heavy oil and its production of one-third of the world's uranium. And, while he covered all bases of non-renewable energy, also mentioning clean coal development, there was no invitation for investment in renewable energy. Calvert's earlier "Green Plan" and the Gull Lake wind farm seemed to be more about optics than energy policy. The emphasis was clearly on oil and nuclear, though a shift in semantics was occurring; the rhetoric no longer spoke of nuclear as a replacement for oil — the CNA's semantic environmental ticket — it was now nuclear as a way to sustain oil.

Calvert wasn't only on a selling spree, he was also shopping for capital. Like his Industry Minister Cline two years before at the World Nuclear Association, Calvert bragged of Saskatchewan's nuclear potential: "As the demand for uranium fuel rises there would be an increased need for uranium refining and he would welcome further private investment in the province."

After his White House meeting Calvert spoke fondly of Cheney, saying, "I found him to be very professional and very welcoming." Like a self-conscious adolescent athlete who had just pleased his coach-mentor, Calvert said:

> The VP was very encouraging about the Saskatchewan resource, the Canadian resource in general, playing an important role in the future energy security of the United States. I think he was interested in some of the information we were able to share about the Saskatchewan resource package, both in terms of crude oil, petroleum products and uranium.[5]

Calvert was so delighted by Cheney's welcoming response to getting more of Canada's resources that he even invited him to come to Saskatchewan. He went to Alberta.

Saskatchewan's corporate media watched Calvert's performance closely, and we can be sure that Calvert's handlers were watching the Premier's "grades." Sounding more like an enforcer of corporate amorality than an investigative journalist, Murray Mandryk of Regina's *Leader Post* said:

> Calvert needs to address the image some have of Saskatchewan as a backwater socialist outpost that the real movers and shakers of the political and business world pass over on their flights to Alberta and B.C. to make deals.... And what better way to offset that left-of-centre image than to rub shoulders with the most hard-line right-winger there is — Dead-Eye Dick Cheney?... if Calvert left the impression that a social democratic Canadian is ready to do business with a U.S. Republican, that's a good thing for Saskatchewan."[6]

With the corporate media and politicians sharing the same unsustainable worldview, it's little surprise that policy-makers and news reporters have become such a mirror image of each other.

Calvert seemed pleasantly surprised at Cheney's know-how on the energy portfolio, noting: "He is no stranger to the energy issues. He knows the file, there's no doubt about that." And why wouldn't he, acting as CEO for the Halliburton Corporation right up until he was elected with Bush in 2000, after which he was the White House point man for the multinational oil industry. Calvert was also notably impressed by the businesslike atmosphere in the White House, comparing it favourably to the television series, "The West Wing." The difference between virtual reality and globalization didn't seem to matter in the compartmentalized world in which Calvert governs, for the star of the West Wing, Martin Sheen, is an ardent Catholic pacifist activist, already arrested sixty-three times, who opposes the militaristic energy policies that Cheney embodies.

Cat's Out of the Bag: Nuclear for the Tar Sands

Calvert's U.S. trip came in the wake of growing pressure from Saskatoon businesses. In February 2005, the Greater Saskatoon Chamber of Commerce passed a general resolution supporting value-added processing of resources. It included "uranium" as one among many examples, which cleverly made economic development, not nuclear energy, the focus of the resolution. Later, its September 20, 2005, newsletter *Business View* had an article on "The Moral Argument for Uranium Processing in Saskatchewan." The moral argument had nothing to do with end uses of uranium and didn't even try to appeal to the CNA's environmental ruse. It simply said: "Securing these activities in Saskatchewan would help keep our kids here [and] could result in several billion dollars of economic opportunity." A notion of "morality" with no recognition of the consequences of one's actions for others or the future is one that has sunk pretty low. Indeed it's "bottom-line amorality."

A year before Calvert's trip to Washington, Dwain Lingenfelter, past NDP deputy leader in Romanow's government, came to speak to a Saskatoon Business Association luncheon to promote Candu reactors and a nuclear waste facility in Saskatchewan. After Lingenfelter left Saskatchewan politics, prior to Romanow resigning as premier in 2001, he moved to Calgary to become the vice president of Nexen Canada Ltd., which is involved in Syncrude and, with offices in Houston and Dallas, is plugged into continentalist energy politics. Like the Devine Tories he helped defeat in 1991, Lingenfelter slammed the Calvert NDP's position on nuclear power for being hypocritical and allowing "mining of uranium for use in reactors throughout the world, but then take a position that it is too dangerous to use for fuel and to deal with the waste locally."[7] He went further, arguing that Saskatchewan should develop the nuclear waste facility that the AECL and Cameco are promoting, adding that Saskatchewan businesses should "become champions... promoting our province as a potential source of clean nuclear power and seeking active investment." (Dwain Lingenfelter along with Jack Messer are a study in how easily Third Way/Tony Blair-like social democrats can bring their insider's know-how from public enterprise into the privatizing, globalizing corporate world. It is to Roy Romanow's credit that upon leaving politics he did something in the public interest, chairing the Federal Commission on Healthcare.)

Lingenfelter reviving the Candu reactor proposal and Calvert pushing a uranium refinery might seem like a two-punch strategy. When a nuclear waste facility for Saskatchewan was raised with Calvert he initially showed interest, but a few days later reversed his position.[8] Was he upholding the Saskatchewan NDP position against nuclear power, or was he learning the political "art" of manufacturing "consent," through incrementalism, for a bigger plan?

Those who see themselves benefitting from Saskatchewan going full-blown into the global nuclear fuel system have been organizing for years with the aid of the CNA and AECL. Calvert's and Lingenfelter's pushes are the latest in a series of attempts to expand the nuclear fuel system in the province. We ought not to forget just how many "plans" the AECL has already thrown at us, from the private Candu reactor plan in 1989, to Tory and then NDP MOUs to develop a Candu-3 industry in the province in 1991–92. The University of Regina re-entered the recent push for nuclear expansion by co-sponsoring, along with the Saskatchewan Urban Municipalities Association and Saskatchewan Association of Rural Municipalities, a conference in January 2006 entitled "Exploring Saskatchewan's Nuclear Future." Even with this title, which assumed there would be nuclear energy in Saskatchewan's future, the conference delegates did not provide a resounding endorsement of such a scenario. If the plans of Calvert and Lingenfelter fail, the nuclear industry will invent new scenarios until it is finally put on the dump-heap of history, where it has long belonged.

In mid-2006, Saskatoon business writer Dwight Percy encouraged those who supported Calvert's uranium refinery to get out and fight for it this time around. He argued the reason the pro-nuclear lobby lost in 1980, when the Blakeney-supported, Warman uranium refinery was turned down, was because it let the non-nukes with their "blowhard anti-nuke rhetoric" win the day. He then advanced his own "boom-town rhetoric," saying that failing to operate a refinery had already lost "billions of dollars" to the province and that a refinery would be "a mechanism by which Saskatchewan can create an image as a place where secondary processing occurs."[9] This figure — billions — is a stretch as this constitutes more than the full value of the uranium industry over this period, most of which goes out of province. Also, like Calvert, who simply talked business with Cheney while ignoring the diseconomics of the nuclear industry, Percy shows no interest whatsoever in the short- or long-term ecological, economic and moral issues involved. When Percy incited the business community to become more activist, he was probably not aware of the larger plan, the one that Lingenfelter was promoting from across the shrinking Alberta-Saskatchewan divide. Realistically, Calvert's uranium refinery probably can't happen unless his and Lingenfelter's scenarios converge. How might this happen? And more vital, what is required to ensure that it doesn't?

Greenprint for Energy Conservation: Too Little Too Late

In May 2006, Calvert put on another hat, a small "green" hat, and appointed long-time Saskatoon MLA Peter Prebble as Legislative Secretary for Renewable Energy Development and Conservation. Prebble's task was to "work with the Energy and Mines Minister Maynard Sonntag to coordinate the development of a Saskatchewan Greenprint for Energy Conservation."[10] Prebble was a non-nuclear environmental activist who turned to electoral politics and was re-elected many times when big name NDPers fell to the right-wing ballot. Mostly marginalized by previous NDP premiers, he was finally in a position to make a difference.

In announcing the move, Calvert emphasized the need to reduce energy consumption as a way to reduce the impact on our pocketbooks. The news release of Prebble's appointment also used Chamber of Commerce-friendly rhetoric about "projects and initiatives to grow the business of developing and commercializing new energy saving technologies." It was noticeable that, along with consumer savings, Prebble but not Calvert emphasized that this initiative would "cut our greenhouse emissions."

The failure to address greenhouse gasses (GHGs) has already caught up with the Calvert NDP. In October 2006, the Suzuki Foundation issued its status report on provincial climate change plans, and not surprisingly, Saskatchewan did not do well. About Saskatchewan, the report said: "GHG emissions are the highest of any province or territory on a per GDP basis.... GHG emissions have grown more since 1990 (62%) than every other province." It noted that the province has "no climate change plan" and continues with its "reliance on coal-fired power."[11]

Perhaps feeling the pressure, Calvert made an important commitment, the first of its kind ever made by a Saskatchewan premier — that by 2020 "one-third of the energy needs of Saskatchewan people will be met by renewable energy sources, and we will lead the country in energy conservation practices."[12] (This promise was to be repeated in the Speech from the Throne in October 2006.) Prebble, always the environmentalist, emphasized,

"Our future, and the future generations to come, depends very much on our stewardship of the environment, particularly with regard to energy.... It is incumbent on us to develop and enhance safe, cost-effective, renewable energy sources...."

Earlier, in January 2006, Prebble had announced he would not be running in the upcoming election. In view of his non-nuclear views on energy policy and Kyoto it was hard to see how he would be able to stay in Cabinet if the Calvert NDP proceeded with plans to build a uranium refinery. Speaking in the Legislature on April 11, Prebble said:

> There is an urgent need to support the Kyoto Protocol and sharply reduce greenhouse gas emissions in Saskatchewan. And that requires a major shift in how we use energy in this province. Mr. Speaker, our policies with regard to the export of energy also need review.[13]

Prebble then went on to express concern about Saskatchewan uranium contributing to proliferation and concluded: "The promotion of a uranium refinery that would facilitate uranium exports should be dropped. Alternative employment should be assured to all those who lose their jobs in the uranium industry as a result of this policy."

Though disengaging from electoral politics, Prebble continued his work on the Greenprint on Renewable Energy. His first report went to Calvert in November 2006 and was made public in December. Entitled *Renewable Energy Development and Conservation: A First Report on Making Saskatchewan a Canadian Leader*, it stressed more investment in conservation and wind power, and new policies supporting net-metering (which allows household renewable electrical generation to be sold back into the grid) and higher energy efficiency standards. Many of these proposals found their way into the government's *Green Strategy*, released in April 2007. A final public report is promised by June 2007, which perhaps coincidentally, is likely to foreshadow a provincial election. It may be "too little too late."

Another Stab at a Uranium Refinery

In June 2006, shortly after launching the Greenprint for Energy Conservation, Calvert — now seeming almost destined to become a tragic political figure — put back on his other, bigger hat, and with Industry Minister Cline went to France to promote investment in a uranium refinery from the state uranium/nuclear conglomerate Areva. Upon their return, they reported that Areva "will look favourably upon Saskatchewan as a location for refining and conversion facilities as the need for increasing capacity arises."[14]

Areva is already involved in refining uranium, which then gets converted and enriched before being used as fuel in France's many nuclear reactors.

(Conversion involves turning a powder [UF4] into a gas [UF6], and enrichment involves increasing the concentration of uranium.) After four years, the spent fuel is removed and sent to Cogema's huge reprocessing plant at La Hague, where some uranium and plutonium are reclaimed for reuse as nuclear fuel, after which, "Casks, weighing 110 tonnes, of spent fuel arrive by boat, rail and truck. Workers use robots of remote control to unload the casks into what look like Olympic swimming pools, where the fuel cools over at least three years…." And then what? Deliberating on "Options for expanding nuclear industry in Saskatchewan," Saskatoon *Star Phoenix* reporter Rod Nichel simply says, without comment, "France's most highly radioactive waste is simply being stored temporarily in the La Hague plants until a permanent solution is found."[15]

Like the U.S. and Ontario Hydro, France has no nuclear waste "disposal" plan, and as the main source of uranium for France's huge nuclear power system, Saskatchewan might be an attractive site for long-term nuclear waste storage, taking back the "rented" uranium fuel, as the AECL and Cameco have put it. Backing a uranium refinery in exchange for a nuclear waste site wouldn't be a bad deal at all, especially if there were also a new market for Areva-mined uranium from Saskatchewan in nuclear power plants in the oil fields, which could provide the rationale for developing a nuclear waste facility nearby. What nuclear country wouldn't jump at such a chance to dump its nuclear wastes and make money at the same time? Invite Japan, or the U.S., which is also having political and ecological problems with its on-again, off-again nuclear waste program, and I think we know the answer. As a competitor with Areva, Cameco might not appear as interested in this, as it already has refining and conversion capacity in Ontario, but throw in a Saskatchewan nuclear waste facility for its Bruce reactors in Ontario, and new demand for its uranium in the oil fields, and the deal would likely be a head-turner.

Far fetched? Perhaps not, for this scenario follows the contours of the integrated uranium/nuclear strategy devised by the AECL in the early 1990s, discussed in chapter 12. The AECL can be accused of squandering public money, which should be going to help us convert to a sustainable society, but it can never be accused of a lack of self-perpetuating self-interest.

The master plan seems to have originated in a paper presented to the Canadian International Petroleum Conference in 2003 that proposed that the heat from nuclear fission be used directly to produce steam for the *in-situ* process used to bring heavy oil to the surface for recovery.[16] This was thought to address the nuclear industry's vulnerability for overall energy inefficiency and high capital and electrical transmission costs, making nuclear competitive with the natural gas presently used to produce steam in the oilfields. However, it failed to consider the thermodynamic stupidity and high costs of transporting steam from the nuclear plant to the oil fields — blindly solving one problem while creating another one of equal or greater magnitude, through the compartmentalized "scientism" that characterizes this industry.

The idea nevertheless continues to percolate through Alberta and Saskatchewan, where the major heavy oil deposits exist. Perhaps Lingenfelter has kept relations with the AECL that were nurtured while in Romanow's cabinet, for he articulated a version of this plan on his 2005 visit to Saskatoon. Another version was announced in August 2006, when the AECL said it had made a deal with recently formed Energy Alberta Corporation to promote a Candu-6 reactor to be built by 2014 to provide steam in the oil fields.[17] Besides targeting companies like EnCana and Husky, this private venture is trying to sell the idea to the oil company, Total SA of France. Maybe it, too, is looking for a way to produce oil with nuclear energy.

This plan appears to be serious, for in the May 7, 2007, *Maclean's*, Wayne Henuset of Energy Alberta Corporation is reported as saying he will file a site application within ninety days for two AECL reactors to be built, starting in 2011 and on-steam by 2016. So here we go again. However, we know well, in Saskatchewan, about these fly-by-night nuclear schemes, and I wouldn't bet my pension cheque on the Alberta plan coming to fruition. The Candu-6, a rebuild of the AECL's older cost-ineffective reactor, is not going to keep the AECL in the highly competitive international reactor market, and the design work for the AECL's new Advanced Candu Reactor won't be done at the earliest until 2010. In the meantime, the advantages of the least-cost alternatives will continue to gain support.

Calvert can't be candid about this larger plan, as he would pay a huge political price for showing interest in nuclear power and a nuclear waste facility. But his trip selling both heavy oil and uranium to Vice President Cheney takes on a new dimension with this AECL nuclear-for-oil plan in mind, which is fundamentally driven by the potential U.S. demand for Alberta's and Saskatchewan's heavy oil. The U.S. is not only the largest per capita consumer of oil and producer of GHGs but the largest importer of oil. By 2025, the U.S. is projected to be importing 93% of its oil. The U.S. is not faring well in its war on Iraq, and its unilateralism and unconditional support for Israel's neo-colonialism in the Middle East is steadily being discredited. An alternative secure oil supply is becoming a strategic consideration. If a second-term Harper Conservative government deepens Canada's military-industrial integration with the U.S., the Canadian Prairies could become, within a decade, the new "Middle East."

The Canadian public needs to enhance our activism for a sustainable future and reject the ecological and socially destructive scenarios being concocted in government cabinets and corporate boardrooms. Something to motivate us is the magnitude in increases in GHGs from ongoing heavy oil ventures. In 1990 the huge mining trucks used in Alberta's tar sands released 50 tonnes of nitrogen oxides per day. By 2003 this was up to 150 tonnes per day, and by 2006 it jumped to 398 tonnes per day. If all the planned $94 billion heavy oil projects proceed, the figure could get to 538 tonnes per day.[18] And 70% of these pollutants will find their way into Saskatchewan, where they will continue to acidify the soil and lakes.

Another way to look at this is in terms of Kyoto targets. Between 2008–12 Canada is to reduce greenhouse gas emissions to 6% below the 1990 levels. In 1990 Canada created 599 million tonnes of CO_2 and other GHGs, so a 6% reduction would bring us to 563 million tonnes by 2012. Yet in 2003 we were already producing 740 million tonnes and are projected to produce 828 million tonnes by 2010. If we stay on our present course of producing non-renewables at the expense of future generations, by 2050 we'd be producing 1,300 million tonnes, or more than double what we've targeted for 2012 under Kyoto. We must also remember that the bitumen in the tar sands has two to three times the carbon as conventional oil, and 80% of Canada's oil would come from the tar sands by 2020.[19] It's clear that the Calvert NDP "vision" of exporting heavy oil to the U.S., with or without using nuclear, is the antithesis of the "green and sustainable economy" he speaks of in his government's *Action Plan*.

The Harper minority government is also caught on these horns of dilemma, for it seems unable to move from its commitment to Canada (Alberta) as "energy superpower," in spite of all the forewarnings about climate change. Remember, Kyoto calls for a 6% cut from 1990 levels by 2012. After political pressure to show that climate change matters politically to climate-change deniers, Harper's government said that by 2020 he would bring levels down 20% from current levels. But current levels are already 27% above 1990 levels, so if we followed Harper, in 2020 our levels would still be 11% above the Kyoto targets set for 2012. I just can't get it out of my mind that Harper's dateline 2020 would allow more nuclear power to get on-stream without interfering with the profitable expansion of the continental oil industry.

The nuclear-for-oil plan has much vulnerability, including its massive capital costs in a market where demand-side management and renewables are more cost-effective. Its greatest vulnerability, however, will come when the Canadian public finally fully connects the dots between energy, environment, sovereignty and war. The environment and non-nuclear movements must diligently work to further this understanding in an open and passionate manner wherever our voices might be heard.

There is another version of the nuclear-for-oil scenario, from the Can West Petroleum Corporation, which would see a giant Candu complex, the biggest in the world, built near Cree Lake, Saskatchewan, 550 km north of Saskatoon.[20] Phase I would take five years and see two 1-GW Candus with a 700 kV transmission line 265 km east to Fort McMurray, Alberta. Phase II would see two more Candu reactors with transmission lines to Calgary, later to join the U.S. grid. The electrical power would be used to create steam for the *in situ* process of extracting bitumen from the tar sands to export to the U.S. It claims that it would replace natural gas, "eliminate" CO_2 emissions and develop Saskatchewan into the largest producer and exporter of electricity in Canada — the latest version of the uranium/nuclear "golden egg." This plan is mostly optics since the uranium and nuclear industries rely on massive fossil fuel inputs, from the mining operations to conversion and the enriching process, which in the U.S. uses electricity from coal-fired plants. The plan amounts to "fossil fuels for nuclear" as much as "nuclear for fossil fuels."

Cree Lake was apparently proposed because "it is the only known place on earth where the natural deposits of oil and uranium exist in such close proximity… [and the] simultaneous exploitation of the uranium and tar sands resources, with nuclear process power producing millions of barrels of oil per day of synthetic crude oil, is an economic roadmap to sustained growth in both industries." This is, unambiguously, a uranium/nuclear vision intent on increasing tar sands production, which by 2004 already constituted 20% of Canada's total output of oil. But the plan was also proposed to address the "Kyoto jitters" rippling through an oil industry that wants to ensure profitable investment to exploit the Athabasca tar sands to meet the growing demand for imported oil in the U.S. At present Alberta produces 30% of the CO_2 in Canada, much of it from the tar sands, and the Kyoto Accord calls for reducing Canada's emissions to 6% below 1990 levels by 2012. So oil-rich Alberta is on the chopping block, or at least behind the eight-ball. A switch from natural gas to nuclear at the tar sands is being pitched because "the nuclear alternative is the only large scale power production possibility that can expand current capacity without CO_2 emissions." This would mean energy growth occurs without proportionate increases in GHG pollution (excluding radioactive pollution, of course); that is, until the oil is used in the U.S., which just happens to not have signed the Kyoto Accord. Again we see the pre-ecological compartmentalized thinking of nuclear "scientism," which ignores the fact that GHGs produced in the U.S. will accentuate global warming.

Nuclear power presently supplies less than 8% of North American primary energy — an amount already surpassed by the renewables biomass, wind and solar energies — without having received billions in subsidies or creating any toxic wastes. So why would anyone propose more nuclear power? Clearly not for itself, except in places like Ontario and France, which have allowed themselves to become so dependent on uneconomic nuclear power that they are having trouble weaning themselves off it. The assumptions of this nuclear-for-oil plan are that oil consumption will continue to rise; that more of it will be imported through to 2025; that nuclear will be more cost-effective than natural gas for heavy oil production; and that changes in culture and technology to reduce oil demand will not occur in this time period. These are all shaky, deeply pessimistic assumptions — but such faulty reasoning has not stopped destructive mega projects in the past.

Regarding the selection of Cree Lake, Saskatchewan, as a possible site, Can West Petroleum's prospectus says: "As a remote and isolated location, Cree Lake would be a preferred place to develop the world's largest nuclear power plant. With a regional population of less than 400 persons within a 200 km radius, many of them employees of the various uranium mining operations, this regional population density is among the lowest found anywhere in the world. No other nuclear plant would have such an extensive natural buffer...." While the plan admits that this location "will add significantly to the cost of an industrial development... what is being proposed here is that the Saskatchewan economy will continue to derive the benefits of uranium production, with the additional secondary benefits of uranium processing, and the tertiary benefits of nuclear power production."

So we can now see how Calvert's push for a uranium refinery could fit into this larger nuclear picture. Perhaps inspired by previous NDP administrations that have tried to spread the nuclear fuel system through covert incrementalism, Calvert may be on his own nuclear learning curve.

The promoters of this plan also pitch security: "This would be the least likely nuclear reactor anywhere to be damaged by 'natural disaster' or to become a 'terrorist target,'" while being so close to the world's largest uranium milling facility. All that is needed is a regional uranium refinery, which Calvert tries to advocate in a precarious vacuum. But the corporate brokers say what Calvert can't say, what is on the tip of the tongue of all the nuclear proponents: "The extremely stable Precambrian rock, coupled with low elevation and basin-like drainage provides an excellent prospect for the permanent deep geological disposal of spent fissile fuel... as well as low-level waste and raw uranium tailings." In a nutshell the plan would pretend to complete the nuclear fuel cycle in one place: "High and low grade nuclear waste could be permanently disposed of at or near the same place it was originally mined as uranium ore.... the world's safest disposal repository could be developed."

This method of "disposal" is what the FEARO panel questioned after eight years of public inquiry, and what FEARO Panel member and past moderator of the United Church, Lois Wilson, discusses in her book, *Nuclear Waste: Exploring the Ethical Dilemmas*. But the AECL, the CNA, Cameco, perhaps Avera and other nuclear proponents can't take "no" for an answer, because the spent fuel is accumulating and threatens to be a long-term corporate cost.

The wastes just won't go away, even if the issues of uranium mining and nuclear power have been made to seem to go away in our present state of collective amnesia.

The Cigar Lake Flood: The Impossible Happens Again

The 2005 *Action Plan* highlights uranium deposits as providing great job and business opportunities, saying "One of the richest uranium mines in the world is gearing up to go into production at Cigar Lake in 2007." Within a year of this pandering to the uranium industry, Cigar Lake's start-up date was postponed to 2009, or even later, after the underground tunnels being prepared to access the high-grade ore flooded and the mine had to be evacuated. The Cigar lake mine has become the idol of the nuclear industry. With its estimated 230 million pounds of uranium, worth $12 billion at today's prices, it creates visions of gold bullion for its owners. It is also held up as a model and considered as a potential site for a nuclear waste repository.[21] So the unexpected flooding of its underground tunnels wasn't very reassuring. If Cameco couldn't get the uranium out without massive flooding, imagine how safe we'd be if nuclear wastes were buried in the abandoned mine shafts.

The first day, October 23, 2006, Cameco reported that a rock fall led to partial flooding, with significant in-flow into a future productive area that was previously dry, but that it would have "no impact on the environment." The company noted: "In accordance with contingency plans, it closed two bulkhead doors to contain the water within the future mining area. Mine shaft #1, the future processing areas, pumps, refuge station and heat exchanger for freezing the ore are being protected from the inflow...."[22] The company was making plans to restore access to the shaft, but admitted construction would be delayed by "at least a year," to 2009, and that the "capital cost is expected to be significantly higher." So, while it was problematic, it wasn't an un-resolvable crisis.

The next day we learned it hadn't gone so well. *Star Phoenix* reporter Murray Lyons wrote that the "immediate future [of Cigar Lake] has been left in doubt," for mine engineers "were unable to get one of the mine's huge bulkhead doors to seal properly and stem the flow of water that has been pouring in at a rate of 1,500 cubic meters an hour since Sunday afternoon."[23] This story was falsely sub-titled "Cameco allows mine to flood" — an interesting way to try to maintain the doctrine of corporate infallibility, whereby the corporation has become exempt from much rational, ethical and ecological criticism and is above "the law," much as the Catholic Church was in the Middle Ages.

The mine had to be evacuated and Cameco was unable to say anything about how it was to be "remediated and put into operation." But we learned more about how the flooding occurred. Drilling within the granite under the rich ore deposit apparently got as close as eleven metres from the extremely porous, water-logged sandstone, which "exerts water pressure as great as ocean water at the same depth." The rock collapsed, Cameco's state-of-the-art contingency plan failed, and "the failure of the gasket within the giant doors, plus the lack of pumping capacity to keep up with the inflow, resulted in the decision to abandon the mine and let it flood."

The Cigar Lake consortium has already invested a half billion dollars into the mine, and according to Lyons, was planning to "employ a method of production using high-pressure water jets to extract the ore and pump it to the surface in a slurry." However, "pieces of its expensive mining equipment that were to be used in the process are now either under water or soon will be." Cameco's CEO, Jerry Grandey, couldn't make "even an educated guess" about a new timetable for production startup. After the failure of the flood containment plan, Cameco stopped further trading of millions more of its shares, which had already dropped in value by 9%.

A far-sighted government wouldn't renew Cameco's licence, issued in 2004 and expiring in 2007. At the time of the flood, its water treatment system wasn't completed and it had no regulatory approval to pump out the mine, and besides, its pumps were below water in the shaft. In these circumstances it's hard to know how Grandey could be so reassuring that, as Lyons reported, "the water is relatively clean… radium levels were low."

It took a *Toronto Star* story to put the flood into a clearer environmental and economic perspective. Apparently this wasn't the first flood that Cameco's mining ventures had caused; there had been a previous flood, though on a smaller scale, at its McArthur River mine site in 2003. And there was also flooding at Cigar Lake in April 2006, which had already pushed back the startup date and raised estimated capital costs from $520 to $660 million.[24] With this new disabling flood, it is clear Cigar Lake will not be able to begin contracted uranium exports in 2008 as planned. This postponement further pushed up the price of uranium: showing that what's good for investors isn't necessarily good for Mother Earth.

Later investigations suggest that Cameco did not have enough pumps and that its blasting may have contributed to the flooding. And this wasn't the first major environmental calamity in northern Saskatchewan brought on by the uranium industry. In 1984 there was a huge radioactive spill at Key Lake, and in 1989 there was another at Wollaston Lake. These and the Cigar Lake flood prove that this industry is not able to operate according to the guarantees given to the inquiries, that this industry cannot be part of Calvert's green and sustainable economy. The Calvert NDP is nothing less than foolish to continue to tie future jobs and opportunities to such an environmentally precarious and dangerous venture as Cigar Lake, when renewable energy alternatives, providing far more and safer jobs that would create a sustainable economy, are so feasible.

The Coming Renewable Energy Economy

Saskatchewan's economy has unfortunately been shaped to depend upon exporting non-renewables like uranium, oil and gas. Because of the blinders of short-term government and corporate self-interest, the province has been slow to realize the potential of renewable, sustainable resources. SaskPower remains far behind public utilities in Manitoba and B.C., where expenditures on conservation (demand-side management) are seen as saving public capital costs for generating electricity. Saskatchewan residents have been misled to think that our economic prosperity depends on the profitability of such uranium/nuclear multinationals as Cameco. And the Calvert NDP has seemingly tied its political future to the expansion of nuclear: to building a uranium refinery in the province.

Meanwhile, worldwide — in spite of huge subsidies to nuclear — renewable energy has now passed nuclear as a source of electricity (20% to 17%). This is partly due to wind, biomass and solar power, but is mostly due to co-generation of electricity from waste heat. Wave (tidal) power will soon accelerate this trend. According to research discussed by Helen Caldicott in *Nuclear Energy Is Not the Answer*, by 2004, decentralized electrical generation provided three times the output and six times the capacity as provided by nuclear, and centralized thermal plants — whether nuclear or coal — can no longer compete with wind and co-generation, and solar may soon become competitive.

Nuclear proponents often reply that "we need all options," but society can't afford all options. And surely we want to encourage the "no carbon" or low carbon alternatives, especially when they don't produce radioactive wastes. Right now the subsidies and hidden costs of uranium/nuclear steal from the potential of converting to renewables. This conversion to renewables is urgently needed to avert the further doubling of greenhouse gasses and perhaps irreversible cataclysmic climate changes over the next half century.

After co-generation, wind is now the "least-cost option" to nuclear power and coal, and the potential for wind power is worldwide. The North Sea, Great Lakes and southern South America are all extremely high potential wind-producing areas. The Canadian Prairies has great potential for wind too. It is good that SaskPower has finally started to build wind farms, but, tied as the government is to the uranium/nuclear industry, it still isn't admitting the great successes of wind power worldwide. According to Caldicott's research, wind power has increased 34% annually since 2001. Denmark rejected nuclear power after Chernobyl and now leads the world in wind technology. England is adding more electricity from wind power than it is losing from shutting down nuclear plants. Germany is nearing 10% of its electricity from wind and biomass.

The huge advantage of wind over nuclear is that it is up and operating much more quickly, with far less energy use and capital expense and no toxic waste, and gives a quick payoff in reductions of GHGs. China, which nuclear proponents assert must replace coal with nuclear, is now exploring wind. It has already constructed a wind farm in Inner Mongolia (at Huitengxile), which by 2008 will generate 400 megawatts (MW) of electricity. The plan of China's Centre for Renewable Energy Development is to have 4,000 MW from wind by 2010 and 20,000 MW by 2020. (20,000 MW is equivalent to ten large nuclear power plants.)

Under George Bush Junior the U.S. government is funding a comeback of nuclear power and has not widely embraced wind. But it is becoming a leader in solar. Sales in solar technology in the U.S. increased by 28% between 2004 and 2005. An estimated 300,000 homes got electricity from solar by 2006. Rebate programs exist in places like New Jersey and California, and forty states now allow residents to sell excess electricity into the public grid.

One downside of solar power is that it takes fossil fuels to create the photovoltaic cells to generate electricity, but solar is still a net energy producer after one to four years. Passive solar, which involves designing and building to capture and retain solar heat, is a completely "no-carbon" option with huge untapped potential. (I know this personally as the super-insulated passive solar house we built in the Qu'Appelle Valley stays at 70 degrees F inside, when it is minus 30 outside, as long as the sun is out, without any other heat source.) The only "downside" to passive solar is that it doesn't produce a commodity for corporations to sell and profit from. It is therefore clearly one of the best sustainable energy options.

North Americans are among the worst energy wasters on the planet. The U.S. Alliance to Save Energy estimates that the U.S. presently wastes half of its energy fuels. And the U.S., with only 4% of the world's population, creates 24% of the GHGs from electrical generation. Meanwhile the Bush administration (along with Canada's Harper minority government) rejected Kyoto and its targets, which could probably be met through conservation and energy efficiency alone.

And Saskatchewan remains part of the problem, with the fastest rising and largest per capita GHGs in the country in recent years. And it remains the major uranium-radioactivity exporting region on the planet. Rather than becoming even more dependent on toxic non-renewables, the Saskatchewan economy needs to shift to a sustainable energy system and economy. Rather than towns being encouraged to compete for a toxic uranium refinery, or nuclear waste dump, or heavy oil plant, Saskatchewan should be promoting a renaissance in conservation and renewables. While ethanol provides a new market for crop farmers, it takes land from food production to fuel machines. Rural redevelopment would be encouraged far more by a growth in "wind farmers." And the province has made some gains in wind power.

If the third (Broadview) wind farm starts up in 2008, wind will provide 5% of the electrical capacity in Saskatchewan. Such sustainable, renewable energy will create far more job opportunities per dollar, without creating long-lived toxic wastes. Policies should be changed so wind farmers can sell excess power into the public grid. It's clearly time for Saskatchewan to get off its destructive and unsustainable energy path.

It isn't only the broader public that is kept in the dark on these positive options. The Saskatchewan non-nuclear movement has itself been mostly asleep since the exhausting days of struggle in the early nineties. Once in awhile it opens a wee slit in one eye to see what's happening, but not for long. Still suffering from exhaustion, demoralization and surplus powerlessness, it readily slumbers off without even telling much of its story to the upcoming generation of activists. Perhaps the speaking tour in Saskatchewan by Helen Caldicott in March 2007 helped "jump-start" a new generation.

The main exception to this prolonged inactivity has been the ecumenical activists in the Inter-Church Uranium Committee (ICUC), who have tried to bring the DU weapons connection and the threat from uranium tailings to the attention of a wider public. For example, in 1999 the ICUC sued the AECB (now the CNSC) for issuing Cogema (Areva) a licence to operate a mill and tailings dump at McLean Lake, arguing that there were serious design flaws not addressed by the environmental review. The ICUC won the first round, lost in a Court of Appeal, was refused standing to appeal at the Supreme Court and is left having to deal with court costs sought by Areva. Without a larger, resilient non-nuclear movement in place, the ICUC has been mostly left to lick its own wounds.

Nevertheless, there may be a clue here to developing the deeper spiritual and communitarian core of resistance, and perhaps activists should seriously consider exploring multi-faith, spiritual forums in their area. Religion and politics are again interrelated — the same fundamentalist religious-political ideologies that are pushing continentalist non-renewable energy systems have helped destabilize Iraq and possibly the entire Middle East.[25] With the new initiatives by the Calvert NDP in concert with uranium multinationals, with the AECL and its radioactive contamination and GHGs threatening the planet, it is time to converse, awaken, meet, rally, protest and, if necessary, civilly disobey the governments and corporations that continue to conspire to undermine our present and future health.

The Calvert NDP may lose the 2007 provincial election, in part because its belated pre-election talk of energy conservation doesn't ring true, especially in light of its persistence with an economic development plan based on exporting more uranium and oil. In the aftermath of several United Nations reports in 2007 discussing the grave consequences of climate change from GHGs, Calvert seems to have shown some change of heart. His government sponsored Al Gore coming to Regina to show the Academy Award winning documentary *An Inconvenient Truth*. Later, in April 2007, his government issued its *Green Strategy*, which discusses how to introduce some renewables, alongside the dominant non-renewable energy systems, within an "energy mix." However, emphasizing clean coal and ethanol while side-stepping the fundamental debate about uranium/nuclear, it's "same old, same old."

The NDP will soon come out with a pre-election *Sustainable Energy Strategy*, but I am not holding my breath for any startling breakthroughs. Actually, rather than tip-toeing obtusely around its energy policy contradictions, the government could study the ideas generated in the clearer-thinking NGO sector, such as in the Saskatchewan Environmental Society's January 2007 *Towards a Sustainable Energy Strategy for Saskatchewan*.

Calvert's NDP will continue to have a tough time trying to put voice to its startling contradictions. It is simply not convincing to say we need more conservation and renewable energy at home, reducing GHGs and energy costs, while at the same time exporting more and more non-renewable oil, which increases GHGs globally. Is the public to believe that while we deserve more clean air and sustainability from renewables, this is not deserved by peoples elsewhere? Is it justifiable to propose uranium as an alternative to oil when the only way the AECL can find inroads to the energy-rich west is to propose a Candu to help produce heavy oil, with all its GHGs? And while uranium/nuclear is not a carbon free energy alternative? And finally, how do you claim to oppose nuclear power here and yet try to justify increasing uranium exports, which spread radioactive contamination around the globe and increase the dangers of nuclear proliferation? You can't, and this moral and ecological contradiction is about to catch up with Calvert and bite him deeply in his political ass.

Creating a sustainable society will not come easily, nor without struggle and sacrifice. But as I have chronicled in this book, there have been many important victories since the CLBI in 1977 — in spite of some lack of coordination and meagre resources. Imagine what the public could do by creating our own community-building transformative processes. Sustainability can be achieved, but by procrastinating transformation we are risking running out of time. And, as I discuss in the final chapter, by continuing to ignore our direct role in escalating the dangers of nuclear weapons proliferation, we will be condemning our grandchildren and their generations of offspring.

Notes

1. "The Other Oil-Rich Province," *Epoch Times*, Mar. 3–9, 2006.
2. Final Statement, World Tribunal on Iraq, Istanbul, Turkey, June 2005, available at <http://www.worldtribunal.org> accessed July 2007.
3. Murray Lyons, "Calvert crossing the Great Divide," Saskatoon *Star Phoenix*, Feb. 10, 2006.

4. Murry Lyons, "Power play: Premier Calvert to talk shop with Vice-President Cheney," Saskatoon *Star Phoenix*, Feb. 10, 2006.
5. James Wood, "Energy pitch to Cheney well-received, premier says," Saskatoon *Star Phoenix*, Feb. 15, 2006.
6. Murray Mandryk, "Did Calvert hit paydirt with White House meeting?" Regina *Leader Post*, Feb. 17, 2006.
7. Abraham Akot, "Lingenfelter touts nuclear waste facility for province," Saskatoon *Star Phoenix*, Nov. 9, 2005.
8. Tim Cook, "Delegates at Sask. NDP convention reject call to prevent uranium refinery," August 8, 2006, available at <www.cp.org> accessed Oct. 2005.
9. Dwight Percy, "Time to show support for uranium refinery," Saskatoon *Star Phoenix*, July 1, 2006.
10. "Greenprint for Energy Conservation to be developed," News Release, Executive Council, May, 26, 2006.
11. *All Over the Map: 2006 Status Report of Provincial Climate Change Plans*, David Suzuki Foundation, 2006, p. 12.
12. "Prebble to Examine Renewable Energy Development and Conservation in Saskatchewan," News Release, Crown Investments Corporation, May 26, 2006.
13. Saskatchewan *Hansard*, April 11, 2006, pp. 1125–26.
14. "Sask. could become home to uranium refinery, government says," *CBC News*, June 28, 2006, available at <http://www.cbc.ca/news/story/2006/06/28/saskatchewan-uranium.html> accessed June 2007.
15. Rod Nickel, "Options for expanding nuclear industry in Saskatchewan," Saskatoon *Star Phoenix*, Mar. 25, 2006, p. F8.
16. See paper by R.B. Dunbar and T.W. Sloan at Can. International Petroleum Conference, Calgary, June 10–12, 2003.
17. Dave Ebner, "Nuclear pitch for oil sands," *Globe and Mail Report on Business*, Aug. 17, 2006.
18. These figures come from a critical discussion of the Harper government's "clean energy" policies during an interview of John Bennett of the Climate Action Network on CBC radio, January 17, 2007.
19. Jeffrey Simpson, "No need for a Kyoto debate: It's over," *Globe and Mail*, Oct. 6, 2006, p. A15.
20. This plan was located at <www.canwestpetroleum.com/s/Home.asp>. Can West Petroleum Corp (now Oilsands Quest <www.oilsandsquest.com/>) is the company developing the heavy oil sands in Saskatchewan.
21. Personal communication, Jamie Kneen, Mining Watch Canada, Oct. 24, 2006.
22. "Canada's Cigar Lake Mine Faces Year-Long Setback After Flooding," NucNet, Oct. 23, 2006.
23. Murray Lyons, "Mine abandoned Monday morning; all workers leave safely," Saskatoon *Star Phoenix*, Oct. 24, 2006.
24. Gary Norris, "Flooding causes delay at Cameco uranium mine," *Toronto Star*, Oct. 26, 2006.
25. See Jim Harding, *After Iraq: War, Imperialism and Democracy*, Fernwood, 2003, which provides a geopolitical analysis that complements what *Canada's Deadly Secret* does on a regional scale.

chapter 18

OVERCOMING AMNESIA

Turning the Page on Uranium Mining

In fall 2006, the Saskatchewan NDP paper, *The Commonwealth*, carried a front page story headed "Sask NDP turns page on uranium."[1] The gist of the article was that because the 2005 convention had (barely) approved the Calvert NDP government's support for a uranium refinery, the burning controversy within the party since 1976 was essentially over. Of course this convention, like past ones, had lots of pro-nuclear maneuvering, which alienated more members, which may come back to haunt the party in the 2007 election. In the article, the government's main "economic proponent" of nuclear expansion, Industry Minister Eric Cline, repeated the mantras that nuclear development presented economic opportunity and that Saskatchewan is only interested in refining uranium for electricity generation purposes. He even asserted that if we don't produce uranium, someone else will, and we should do it because "we know it will be done right." I suspect the Industry Minister and the Premier believe this rationale, at least as much as they are able. It is politically understandable that Cline hopes he is right when, regarding this issue, he says, "I think both the public and the party have moved on." But he and Calvert, and Calvert's Saskatchewan Party opponent, premier-in-waiting Brad Wall, are going to find out that you can't really turn a page, or move on, when the issue is the slow, steady, deadly radioactive contamination of ecosystems and people and the continuing threat of nuclear conflagration.

If we are going to turn a page we probably ought to turn it backwards, then take a deep breath and think carefully, with an open heart, before beginning again. Encouraging public amnesia and making false reassurances to keep this deadly secret doesn't alter the fact that we are up to our ankles in radioactive uranium tailings, that we continue to contribute to nuclear proliferation and that people are dying from the depleted uranium (DU) bombs created, with great help, from Saskatchewan uranium.

We have all experienced how difficult it can be to admit, on a personal level, that our actions are causing problems for others. How much more difficult must it be, then, for a figurehead or political representative or spokesperson to do so. Mental compartmentalization, along with a touch of executive power, may help us hold on to our delusion, but being keepers of deadly secrets that threaten our wellbeing and the wellbeing of others is not something we should wish to hold on to.

If we are going to eliminate nuclear weapons of mass destruction (WMD) and the threat they pose to the evolution of life on this planet, if we are to make a successful transition to sustainable energy technologies and international law, we must fully lift the veils that collectively blind us to the history and the persistence of these monstrous inventions. Politicians in Saskatchewan, as elsewhere, are adept at pointing fingers at others for the threat of nuclear weapons. We see this in its most crass form with the increasingly discredited Bush Junior regime, where "WMD" became a political football in the "single superpower" geopolitical climate following the collapse of the Soviet Union and the post-9/11 manipulation of people's deepest fears about Iraq having WMD — a smokescreen for the U.S.'s illegal invasion and occupation. A similar tactic is also being tried with Iran.

It is self-serving and deceptive to displace the responsibility for nuclear proliferation when the production of these weapons has become deeply ingrained in the global military-industrial system, of which Canada is a part. Technologies in hinterland as well as metropolitan economies are all being shaped by this, and there are many places on the planet where the veils of secrecy have to be lifted. As the northern neighbour to the world's largest military-industrial economy, Canada may have the most to reveal. Canada's branch-plant economy and our role as a continental and imperial resource hinterland has steadily pushed us in the direction of military-industrial integration.[2] While Canadians want to be seen as "peacekeepers," and Canada has a somewhat independent tradition in foreign policy (Suez, Vietnam and Iraq), it has been deeply complicit in the creation and proliferation of nuclear WMD.

There is a long litany of examples of Canada's direct role in this: from the Manhattan Project in World War II; to the first nuclear arms race of the 1950s–60s; to the spreading of Candu technology that facilitated proliferation; to the second nuclear arms race of the 1980s; to the creation of a new line of radioactive DU weapons used in the Gulf War and the invasion of Iraq. Canada has been involved throughout in laying the ground for mass death and ecocide by nuclear means. And, as Saskatchewan's uranium mining has become the major front-end in the global nuclear system, its role in nuclear weaponry has become more central.

Though active in the non-nuclear movement for decades, I didn't realize the extent of our complicity until I prepared to present at a roundtable on "The Preconditions for the Elimination of Weapons of Mass Destruction," on November 22, 2004, at the University of Regina. Initially, I felt like this topic should be "taboo," as it brought to the surface many fears that have been with me since I became active as a teenage Ban-the-Bomber. Until I sat down to prepare this difficult "subject matter," I had not connected all the dots on what I knew until then only in compartments. All the presentations were edited and published in 2006 by my past colleague in the School of Human Justice, Yussuf Kly.[3] So, before I end this life-encompassing book, I want to connect these dots for the reader by doing a detailed survey on just how complicit Canada and Saskatchewan are in this deadly nuclear history. The historical information on Saskatchewan from 1953 to the Blakeney premiership is based on a report by myself and Dave Gullickson.[4]

From A-bombing Hiroshima to Contaminating Iraq with DU

Canada's involvement began during World War II, when Eldorado's Port Radium mine in the N.W.T. was reopened in 1942 for the purpose of mining uranium for the Manhattan Project. By 1943 Canada had become a full-fledged participant because of its fundamental importance to the Project, with Canada being a major source of uranium for the atomic bombs dropped on Japanese civilians at the end of the war. Eldorado was made into a federal Crown corporation in 1944. Thus began the long history of keeping the deadly secret of the irrevocable connection between state-based nuclear reactors and nuclear weapons.

Nuclear "development" occurred without any popular knowledge or democratic accountability. For twenty years, from 1942 to 1962, Canadians had no glimmer of our government's central involvement in the creation of the first nuclear weapons of mass destruction. As Carole Giangrande pointed out, Eldorado's 1962 Annual Report finally admitted that, "Canada's role [in the Manhattan Project] was to supply the uranium raw material."[5] What they never admitted was that it was uranium from Port Radium that was used in the bomb that destroyed and radiated Hiroshima.

Canada hasn't only been a resource hinterland for nuclear weaponry, it also played a central technical role in the creation of the first nuclear weapons arsenal. The experimental nuclear reactor constructed in 1945 at Chalk River, Ontario, was the first reactor outside the U.S. used to create weapons-grade plutonium for the U.S. While at the war's end the Canadian government told the Canadian public it was only developing nuclear technology for non-military industrial purposes, as Ronald Babin pointed out, it continued to directly supply the U.S. military with plutonium. Babin noted: "This activity increased in scale as more reactors were built in Canada, and continued until 1963."[6]

These experimental reactors were the first to use heavy water and led to the development of the Candu nuclear technology, which came to play its own role in nuclear proliferation. While the Candu is constantly marketed as a peaceful use of the atom, it is vital to remember that, as Babin showed, "Canada began the development of military and civilian nuclear technology as two products of a single scientific effort," and such interconnections continue to this day.

The uranium industry in Canada expanded along with the proliferation of nuclear weapons. While the U.S. had some domestic uranium supplies, it became increasingly dependent on Canadian sources as the Cold War escalated. Between 1942 and 1954 about 30% of the uranium required for U.S. nuclear weapons' production came from Canada.

Under the leadership of Tommy Douglas, the Cooperative Commonwealth Federation (CCF) was first elected in Saskatchewan in early 1944. At about the same time, so it could control production of uranium for bombs, the Canadian government used the *War Measures Act* to expropriate Eldorado Gold Mines. In 1946 the new Crown, Eldorado Nuclear, staked its chief uranium claims in the Beaverlodge area in northern Saskatchewan, and the CCF initiated the Prospectors Assistance Plan to encourage uranium development and requested, but failed to get, a seat on the Atomic Energy Control Board in Ottawa.

Soon after, the head of Saskatchewan's new Economic and Planning Board explored the possibility of the British Ministry of Supply mining the province's uranium deposits, but the British indicated they had sufficient sources for their weapons program. In 1951, when the U.S. Atomic Energy Commission (USAEC) guaranteed a market and a profitable price for all Eldorado uranium production, Saskatchewan became a major supplier of uranium for the U.S. weapons program. The other main source was Elliot Lake, Ontario, which by 1960 had eleven operating uranium mines. Several other mines opened after the Beaverlodge mine started production in 1953, and by 1958, 5,500 people lived at Uranium City. Saskatchewan uranium production peaked in 1958, when six million pounds of uranium (30% of Saskatchewan's mineral production that year) was mined and exported for nuclear weapons. However, purchases were guaranteed with the USAEC until 1962–63. Between 1956 and 1963, thirty million pounds of uranium were exported from Saskatchewan to the U.S. for weapons' production, which was more than was exported until the uranium "boom" in the late 1970s.

The Lorado mine closed in 1960 and the Gunnar mine closed in 1964, leaving only Beaverlodge in production. The expectations of continued demand for uranium for nuclear-generated electricity — after President Eisenhower's "Atoms for Peace" speech in 1953 during the Korean War — were greatly inflated. However, demand for uranium continued due to United Kingdom Atomic Energy Agency weapons contracts after 1957 until 1967, and extended to 1971.

Until the 1970s, 90% of uranium from Saskatchewan went directly to the U.S., and, as there was no commercial nuclear industry until the mid-1960s, we know that all this uranium went directly into military uses. It is estimated by Giangrande that, overall, Canada supplied the uranium for one-third of the 26,000 nuclear weapons in the U.S. arsenal during the post-World War II period.

One of the biggest political ironies of my life was inviting my teenage hero, the late Premier Tommy Douglas, to speak at one of our largest nuclear disarmament rallies at the Saskatchewan Legislature in the late1950s, not knowing that at the time he knew our uranium exports were for U.S. and U.K. nuclear weapons' production. Before his death he told me he was sworn to secrecy under the *War Measures Act*. (I vigorously discussed this with Tommy Douglas in the late 1980s, at a conference at Fort San, and believe some of this is on film at Birdsong Film in Regina.) It is little wonder Tommy wanted his legacy to be attached to memories of Medicare and not memories of exporting uranium for nuclear weapons.

As premier, Tommy Douglas covertly supported uranium mining that provided fuel for the buildup of nuclear weapons. But when in 1963 the Pearson Liberal government did an about-turn and supported bringing nuclear-armed Bomarc missiles onto Canadian soil, Douglas, now a Member of Parliament, was outraged and supported the Ban-the-Bomb movement. This split between helping to produce nuclear weapons but not being willing to house them, is somewhat like the contemporary Saskatchewan NDP's split: agreeing to export uranium for nuclear energy elsewhere but not wanting to have a nuclear reactor "at home." The big difference is that Douglas finally admitted his government helped fuel nuclear weapons, whereas today's NDP leadership is still in denial about their own complicity.

Saskatchewan persisted down the nuclear path after the first nuclear arms race "quieted down" with the Limited Test Ban Treaty of 1963. And prior to 1976 much of this quest for "nuclear development" occurred completely out of public view. In 1965 the Thatcher Liberal government attempted but failed to get an AECL heavy water plant at the town of Estevan. Soon after, corporations from several countries with military-industrial nuclear programs became active in uranium exploration in the Saskatchewan's north. In 1967 the French consortium, Amok, began aerial surveying, and in 1971 the Cluff Lake uranium deposits were discovered. In 1968 Gulf Minerals discovered a uranium deposit at Rabbit Lake, near Wollaston Lake, and in 1969 the German Corporation, Uranerz, joined in the Gulf Minerals' venture and guaranteed a market for all Rabbit Lake production.

In 1969 the Thatcher Liberal government began building roads to the Rabbit Lake and Cluff Lake sites. Later the Blakeney NDP government agreed to provide social capital for the Collins Bay settlement, at the Rabbit Lake mine site, though the uranium companies soon decided to develop a commuter system for workers rather than a company town. In 1971 the Saskatchewan government tried, but failed, to get a uranium enrichment facility at Estevan and even considered building a new coal-fired electrical plant to power it and to divert the South Saskatchewan River for the astronomical amount of water that would have been required. In 1973 the NDP government lobbied the federal government for an AECL heavy water plant to be built at Estevan, but this too failed. The diversion of massive amounts of water from Lake Diefenbaker was considered for this venture.

After the price of uranium collapsed in 1971, a uranium cartel, including Canada, operating from 1972–74, tried to keep the price of uranium artificially high. In 1973 OPEC raised the price of crude oil and the "energy crisis" was on. It was within this climate that in 1974 the Blakeney NDP government revised its minerals policy and created the Crown corporation, the Saskatchewan Mining and Development Corporation (SMDC) to form joint ventures with uranium multinationals.

Eldorado Nuclear set up its Port Hope, Ontario, refinery soon after it established the Port Radium mine in the N.W.T. While it was initially used to refine radium, after the Manhattan Project began, it refined uranium that went to the U.S. for weapons' production.

In 1976 when Canadian uranium production was shifting to Saskatchewan, Eldorado proposed a new uranium refinery, to be closer to the new, large Cluff Lake and Key Lake mines. The preferred site was Warman, near Saskatoon, with access to railways going east and west and truck routes going south to the U.S. In 1980 federal hearings were held on the proposed new refinery, which the Blakeney NDP government backed, but opposition throughout Saskatchewan was overwhelming. Opposition was steadfast within the local pacifist Mennonite community targeted for the refinery, and the board of review recommended against the location because of the potential social impact. Its report stated, "The panel recognizes… the concerns about proliferation of nuclear weapons. Because of the extent of public concern, it believes that the federal government should continue to pursue institutional means to strengthen international safeguards."[7] Eldorado later established a new uranium refinery at Blind River, Ontario, close to the Elliot Lake uranium mines, which were soon to become uncompetitive with the Saskatchewan mines, leading to massive layoffs from 1990 to 1996.

Canadian nuclear technology directly contributed to the nuclear arms race in Southeast Asia and almost played a role in nuclear proliferation in South America and the Middle East. Canada supplied India with the Candu technology in 1972 and a little later supplied it to Pakistan. Wanting "nuclear parity" with China, India used this technology to help produce its first nuclear weapon in 1974. When India exploded its first atomic bomb, Pakistan felt threatened and wanted "nuclear parity" with India, and it soon produced the first "Islamic bomb." Though Canada, after considerable hesitancy, officially banned fuel or reactor parts to India and Pakistan, the regional arms race was on. India and Pakistan, along with Israel, remain the only counties that still have not signed the non-proliferation treaty. Still unknown to many, North Korea joined the NPT in 1985.

Canada didn't learn the hard lesson: later South Korea was the recipient of a Canadian research reactor and then a Candu, just when it was also exploring nuclear weapons' production. The SMDC (now Cameco) was also negotiating uranium exports to South Korea at the time. In 1973, the AECL started negotiating a Candu sale to Argentina's military dictatorship. In 1981 Argentina's new government admitted its former nuclear program was for military and strategic, not commercial, reasons. Had the Candu sale been made at that time it might have triggered a nuclear arms race with Brazil. A Candu was ultimately sold to Argentina and became operational in 1984.

We have already seen that the AECL's sales reps apparently have no scruples; in 1974 they even attempted to sell a Candu to Saddam Hussein's Iraq, with one main selling point being the reactor's capacity to produce plutonium.[8] In 1980 the AECL made its largest Candu sale ever (five reactors) to Romania, while it was still under authoritarian rule. These reactors were built sub-standard and drained scarce resources, seriously inhibiting the desperate country from meeting its people's basic human needs.

Though the AECL and federal and provincial governments have continued to hide behind pronouncements that appear to honour the NPT, Candu exports have clearly contributed to nuclear proliferation. Official statements to the contrary, there have been other ways that Canada has breached the NPT.

After the discovery here of the most concentrated uranium-bearing ore in the world, Saskatchewan moved to the centre of Canada's involvement in nuclear proliferation. The first high-grade uranium mine opened at Cluff Lake in 1977 and was operated by the French company Amok, with the SMDC owning 20%. Amok was connected to the highly centralized French Atomic Energy Commission, which oversees both the state-run nuclear power plants and the nuclear weapons program.

Canada approved contracts for Amok and the SMDC to sell to France, even though France operated its commercial and weapons program under the same authority and had not signed the 1970 NPT. (France, along with China, did not sign the treaty until 1992.) At the time France was still carrying out above-ground nuclear weapons' tests in the South Pacific. Amok also sold uranium to Germany, which in turn sold it to Brazil, which at the time was threatening to develop nuclear weapons in reaction to Argentina's nuclear program.

The Saskatchewan government's own environmental review body finally admitted the weapons' connection. Referring to the French company, Cogema's, proposed mine at Midwest Lake in northern Saskatchewan, the 1993 report of the Joint Federal-Provincial Panel (JFPP) on Uranium Mining said: "specific proponents, such as Cogema, are wholly-owned subsidiaries of foreign governments heavily involved in weapons research, fabrication and testing."

Canadian uranium exports to the U.S. were stopped in the early 1970s, due to court action by U.S. nuclear power companies over the uranium cartel, in which Canada was complicit, artificially keeping uranium prices up.[9] At the time there was also a growing protectionist U.S. uranium lobby, which felt threatened by Saskatchewan's huge new, lower-cost uranium mines. However, by 1975, Canadian uranium imports were allowed into the U.S., and by 1979, one-half of Saskatchewan's uranium went straight south, where it was supposedly only enriched for light water reactors, not for nuclear weapons' production.

The Canadian and Saskatchewan governments claimed that this burgeoning demand for Saskatchewan uranium was due to the expansion of nuclear power plants. However, though there were grandiose claims about the inevitable expansion of nuclear power after the OPEC oil price increases in 1973, the expansion didn't materialize, especially after the Three Mile Island near-nuclear meltdown in 1979 and the Chernobyl catastrophe in 1986. There was, however, growing demand for uranium for the second and even larger nuclear arms race, which took place throughout the 1980s.

The Reagan administration steadfastly opposed U.S. protectionism and supported "free trade" with Canada to ensure security of uranium supply. And the uranium industry operating in Canada was among the earliest backers of the Free Trade Agreement. Under Reagan's presidency, 37,000 more nuclear weapons were produced, which was far more than those produced from 1945 to his presidency. The extreme economic pressure created by this massive nuclear arms buildup, as well as the Chernobyl nuclear crisis, contributed to the collapse of the Soviet Union.

Once the U.S. border reopened to uranium from Canada, military reactors that had been shut down in Southern Carolina coincidentally came back into operation. Giangrande pointed out that even when uranium supply glutted the commercial reactor market, the U.S. continued to stockpile a further 90,000 tonnes of uranium, much of it from Saskatchewan.

The Saskatchewan and Canadian governments engaged in legalistic hair-splitting, arguing that the actual molecules from Saskatchewan uranium did not go directly into U.S. weapons. However, similar to France, the U.S. Department of Energy oversees uranium enrichment for both nuclear power and nuclear weapons' production. And, as Giangrande said, "It's hard to see how the U.S. could have maintained both a civilian and military nuclear programme without importing some Canadian uranium."

The 1993 JFPP on uranium mining, which acted on behalf of the Romanow as well as the federal government, felt it necessary to highlight the matter of the end uses of Saskatchewan uranium: "There is no process whereby exported Canadian uranium can be separated from uranium derived from other sources. Therefore no proven method exists for preventing incorporation of Canadian uranium into military applications.... Current Canadian limitations on end uses of uranium provide no reassurance to the public that Canadian uranium is used solely for non-military applications by purchasers." The World Commission on Environment and Development, or Brundtland Report, came to a similar conclusion in 1987.

So, there it is, the CCF-NDP governments, along with various federal governments over more than a half century, have been directly complicit in nuclear weapons proliferation. And once you put all this history together, it becomes tempting to rename the Saskatchewan NDP the "Nuclear Development Party." While it's hard to break the mould of denial, with so much collective amnesia in place, it simply must be done. And it must be done soon, since Saskatchewan uranium is going to continue to contaminate life until it is stopped "at the source."

Tracing Saskatchewan Uranium into DU Weaponry

In his March 25, 2006, story, "Sask uranium used for energy, not bombs," Saskatoon *Star Phoenix* writer Rod Nickel asked: "Is our uranium being used in weapons?" and then quickly answered: "The short answer is no. Uranium produced in Saskatchewan is not being used for weapons, at least not legally or within detection of the IAEA monitors." When he asked Saskatchewan-uranium importer Areva about the concern about fuel reprocessing — the French practice of recovering plutonium from spent fuel, which past U.S. President Jimmy Carter got the U.S. out of because of concerns about nuclear proliferation — Nickel was told that this "degrades plutonium making it unsuitable for weapons"; and anyway, Areva only enriches to 4–5% "not close to the 90% level needed to manufacture weapons." With this reassurance and after getting International Atomic Energy Agency guarantees that Saskatchewan uranium isn't diverted for military purposes, Nickel seemed satisfied.

Avera and the IAEA reassurances are challengeable — and outdated. The way Saskatchewan uranium contributes to radioactive weaponry took a new turn with the invention of depleted uranium (DU) bullets. It is vital to trace how this makes Saskatchewan's uranium industry and government complicit in waging what can be called a low level nuclear war.

How did these DU weapons originate, and what evidence is there that Saskatchewan uranium exports are involved? It turns out that over the first fifty years of enriching uranium for nuclear weapons and commercial reactors, the U.S. accumulated about 500,000 tonnes of DU (U-238),[10] and some clever arms industry researchers invented a means to use this to make a profitable military product. This can be seen as the military-industrial complex's perverted version of recycling. Because uranium is the heaviest element on the planet, the arms researchers came up with the not so complicated idea that bombs containing uranium could be used to pierce less dense metals, like the steel casing on tanks. It was a classically destructive engineering feat, with no consideration of what happens when a DU bullet hits a target and disperses radiation on the ground and into the atmosphere.

These DU weapons were first envisaged during the Manhattan Project in 1943. They were provided to Israel in 1973, when it went to war with several Arab states. But as Leuren Moret wrote, they weren't mass-produced and used until 1991, in the Gulf War.[11]

The Canadian and Saskatchewan governments are either extremely naive or deliberately deceptive when they guarantee a questioning public that our uranium no longer goes into military uses. Though there may have been "stops" to using our uranium directly in nuclear weapons, after contracts for weapons purposes ended in the late 1960s, DU left from these and recent sales is part of the accumulated stocks now being used for producing bullets.

After imported yellowcake is enriched by the U.S. Department of Energy, the remaining DU is treated as "U.S. material," and this is true whether it comes from the U.S., Australia, Namibia or Canada. Considering the estimate that Canadian uranium was used to create one-third of the U.S. nuclear arsenal up until the late 1960s, we can assume that our uranium contributed one-third of the DU stocks coming from that weapons' production.

But it doesn't end there. After Saskatchewan's high-grade uranium mines began production in the mid-1970s and the U.S. had shut down its own mines and targeted Canada for security of uranium supply under the FTA, Saskatchewan contributed a much higher percentage of the stockpile from which DU weapons are made. Even if the average of about 4,000 tonnes of yellowcake, exported from Saskatchewan to the U.S. each year since the early 1990s, was initially used only for fuel for nuclear power and to free other sources for weapons' production, the depleted uranium from this remains in the "U.S. material" drawn upon for DU weapons' production.

Of the estimated 500,000 tonnes of DU that the U.S. has accumulated since the Manhattan Project, exactly what percentage is from Canada and Saskatchewan cannot be fully determined. But regardless of the percentage, there is no doubt our contribution is significant. Through the 1950s and 1960s, and even more so since the 1970s, as the world's largest uranium-producing region and the U.S.'s major source of supply, we are no doubt right at the top of complicity in DU weapons' production.

Cameco's records show that from 1991, when it became a publicly trading corporation, until 2005, it produced nearly 260 million pounds of uranium.[12] And by 2010, if Cigar Lake comes on stream, it expects to produce 30 million pounds a year.[13] The percentage of sales from this production to the

U.S. is hard to calculate. Though Cameco's contracts remain confidential, we however know that 60–70% went to all of North and South America between 2001 and 2005. With 103 reactors in the U.S., compared to Canada's thirty, and only a handful in South America, it's not unreasonable to say that a half of Cameco's exports go to the U.S.

That would mean about 130 million pounds (or about 65,000 tons) of uranium going to the U.S. from Saskatchewan's Cameco production alone in the last fifteen years. And more might go to the U.S. from other sources, which Cameco buys from to fulfill contracts. Ninety percent of all this would end up as DU, contributing to the stockpiles used for DU and other nuclear weaponry.

But this only begins to account for uranium exports and DU contributions from Canada and Saskatchewan during the longer period, from 1953 to 2005. Between 1956 and 1963 we know that 30 million pounds (or 15,000 tons) of uranium went directly to the U.S. from Saskatchewan, and likely much more went to the U.S. from the Elliot Lake, Ontario mines. When I looked at Saskatchewan uranium production volume, back to 1963, which is as far as public records now go, I found that from 1963 to 2005, 609 million pounds of uranium had been produced in Saskatchewan. (I hope the missing data from 1953 to 1963 isn't our attempt to erase all memory of our uranium-military contracts with the U.S. during the 1950s and 1960s.)[14] This converts to about 304,500 tons. If we assume half went to the U.S., then this totals 152,250 tons. Adding the 30,000 million pounds (15,000 tons) from the period 1956–66, we get about 167,265 tons from Saskatchewan alone, which would create a large chunk of the DU stockpiles left in the U.S. When DU from Elliot Lake mines is also considered, it is not unreasonable to assume that upwards of one-third and perhaps more of the U.S. stockpile used for DU and nuclear weaponry comes from Canada, with most of this from Saskatchewan due to the high exports in recent decades.

The production of DU weaponry is a profitable business, and the U.S. has several factories producing these weapons. They are also produced in the U.K., France, Russia and Pakistan. DU ammunition is now part of the arsenal used on A-10 Warthog aircraft, AH-64 Apache helicopters, Abram tanks, British Challenger tanks, Bradley Armoured vehicles, Phalanx guns on naval vessels, Vulcan and Avenger Cannons, and Tomahawk Missiles. Selling DU weaponry has become a worldwide business, with these deadly products already going to twenty-nine other countries.[15]

So DU from Saskatchewan uranium is certainly getting spread around the globe. And one of our legacies as the major uranium-producing region on the planet is helping to contaminate several war zones with long-acting radioactivity, which is contributing to civilian deaths long after the military conflicts have ended.

We don't require direct evidence of the weapons' connection to conclude that Saskatchewan uranium exports breach the NPT by contributing to military production using nuclear materials. Nevertheless such direct evidence exists. In 1993 the Saskatoon-based Inter-Church Uranium Committee (ICUC) released copies of a licence from the U.S. Nuclear Regulatory Commission that suggests that DU from Saskatchewan uranium was directly used for producing these weapons. As much as I am able to discern, the steps in this process were as follows:

- Yellowcake was exported from the SMDC (now Cameco) to the Sequoyah Fuels Uranium Conversion Facility for refining. (In 1986, 480,000 pounds was shipped and shipments continued until 1992.)[16]
- Sequoyah Fuels then refined the yellowcake into a) uranium hexafluoride (UHF) for reactor fuels, leaving b) depleted uranium tetraflouride (UF4 or DU).
- Sequoyah Fuels then supplied DU to Aerojet Ordnance Tennessee (AOT).
- The AOT had a licence from the U.S. NRC to send up to a million pounds of DU to Eldorado Nuclear's (now Cameco's) uranium refinery in Port Hope for production into uranium metal. The licence was for 1988–90.
- The purpose of the uranium metal was to make armour-piercing munitions ("DU penetrators") for the US military.[17]

That the Port Hope refinery was used to create the material for DU bullets seems unquestionable, even if we wish to debate whether the DU originated in Saskatchewan, for the NRC licence says: "AOT will supply the UF4 to Eldorado Resources who will use it to manufacture Depleted Uranium metal for AOT's use in the manufacturing of depleted uranium penetrators on U.S. Dept. of Defense Contracts." And certainly we can debate whether this practice has been stopped or not.[18]

But even this astonishing evidence doesn't begin to tell the full extent of the use of Saskatchewan uranium in weapons' production. We have already seen that the only safeguard against uranium being used for weapons is the NPT, signed in 1970. And this treaty is what NDP Industry Minister Cline, Premier Calvert and Cameco officials appeal to in their public claims that we are only producing uranium for peaceful uses. However, while this treaty, following on the Test Ban Treaty, was an important step forward, it is not based on a solid understanding of how the nuclear fuel system and the weapons-making process works.

The head of the Canadian Coalition for Nuclear Responsibility, Gordon

Edwards, reminds us: "To produce just 1 kg of 5% enriched uranium requires an input of over 11.8 kg of natural uranium, and results in 10.8 kg of depleted uranium (having about 0.3 % U-235)."[19] He continues: "In other words, over 90% of all Saskatchewan uranium that was ever sent to the USA for enrichment (for peaceful purposes as nuclear reactor fuel) has remained in the USA as depleted uranium (DU)." He further reminds us:

> There is absolutely no distinction between the DU of Canadian origin [or Saskatchewan origin] and the DU of other origins (US, Australia, etc.) It all goes into the same large stockpiles of DU. [And...] a portion of this large stockpile of DU has always been used freely and without any compunctions by the US military for military purposes.... Thus there is some Canadian DU in every DU weapon.

There is even more to the contemporary weapons connection: in spite of all the reassurances that our exports are only for "peaceful purposes," the DU from the common stockpile is also used directly in the construction of the metal warheads in H-bombs. Edwards notes: "This depleted uranium is responsible for at least 50% of the explosive power of each H-bomb and almost all of the radioactive fallout from the H-bombs." And he continues: "Most people do not realize that the same DU stockpile was also used for half a century and more to produce the plutonium that is used in almost all nuclear warheads."

All the reassurances based on the NPT collapse with this specific knowledge of how the weapons-producing process actually works. Though our collective conscience may have been eased a little when Saskatchewan uranium contracts explicitly signed for nuclear weapons ended in the late sixties, our exports from then to now have continued to directly contribute to nuclear weaponry.

And some of this weaponry has already been used. DU weapons were used in the 1991 Gulf War, in the Balkans and Kosovo between 1992 and 1995, in Afghanistan in 2001 and in the invasion of Iraq in 2003. The tens of thousands of rounds used in the first Gulf War probably created somewhere in the range of 30–50 tonnes of radioactive and toxic waste on this fertile land. (The U.K.'s Atomic Energy Authority says at least 36 tonnes whereas the IAEA estimates 50.) Childhood leukemia and some other cancer rates are already five-fold what they were in Iraq before the Gulf War, and the effects from the 2003 invasion and occupation have only begun to show. I heard medical evidence of this presented at the World Tribunal on Iraq in Turkey in June 2005. Comparative epidemiological research is ongoing.[20]

The after-effects of the DU will be far more devastating to human life than the actual firing of the weapons. In some bombed areas radioactivity became from 300 to 1,300 times the background level. Iraqi children have been found dismantling ordnance for metal scraps. In one survey birth defects such as no limbs or eyes or other deformities were up eighteen-fold in 2001 over 1990. Though the numbers from one study are small (from 37 in 1990 to 611 in 2001), the rate of increase is profound and the full effects, including from 2003, will not yet be apparent. Even so, U.S. troops have not undertaken decontamination procedures, and at the time of this writing the Pentagon would still not release information to the United Nations Environment Programme and the World Health Organization on where these DU bombs were used.

DU radioactive contamination puts soldiers as well as civilians at risk. Dutch troops were sent into the Al Muthanna area in Iraq without being told that DU weapons had been used there. There are already signs that offspring of soldiers exposed to DU in 1991 are more at risk of such conditions as heart valve, urinary and kidney defects. Many U.S. veterans suffer from Gulf War Syndrome, which is thought to involve exposure to DU weapons.

Low Level Nuclear War: Happening Now

The scale of use of DU weapons and geographic range of the radioactive pollution are massive. Fifteen hundred bombs were dropped on Baghdad during the first twenty-four hours of the "Shock and Awe" invasion of Iraq in March 2003. An estimated 300,000 rounds of weapons containing DU were fired from A10 Tankbuster planes during this assault. Seven to nine days later, the measurements of uranium aerosols in the air at Aldermaston, England, increased four-fold, twice exceeding the threshold that is required to inform the U.K. Environmental Agency. These were the highest levels of uranium particles ever measured in Britain.[21]

The official interpretation was that it was strictly coincidental that this occurred soon after Shock and Awe and that it must have been local contamination from the Aldermaston weapons plant. This would be quite an admission from nuclear officials. However, if this were true, there would have been fission materials like plutonium in the filters, and there were not. The back-up official explanation, totally contradicting the first, was that the unimaginable firepower of Shock and Awe had stirred up natural uranium in Iraqi soil, which had found its way to Britain. However, uranium in the earth's crust is about 2.4 parts per million, and this would dilute rather than concentrate to the higher measured British levels during the 2,400 mile journey from Iraq. Besides, there are no known uranium deposits in Iraq.

The non-official view is that this radiation came from the enormous amounts of DU weaponry used by the U.S. and U.K. troops in the invasion of Iraq. This interpretation continued to gain support when the official explanations became self-contradictory and it was found that wind currents and airflow during that time were actually from Iraq towards Britain.

The public, however, would know none of this if environmental scientist Chris Busby hadn't used freedom-of-information legislation to force officials to release the radiation data. At first, the Atomic Weapons Establishment (AWE), which did the measurements, refused all access to the data. (The AWE began to collect this radiation data in 1991, after a cluster of childhood leukemia was discovered in the area around the weapons plant.) When Halliburton, which took over the AWE, was finally ordered to release the information, the 2003 data were missing, and Busby had to go to the Defence Procurement Agency to get the measurements. Busby has impressive qualifications for assessing the significance of these extraordinarily high uranium readings. The founder of Green Audit and an international expert on low-level radiation, he recently co-authored the European Committee on Radiation Risk Report (ECRR) for the European Parliament. It fundamentally challenged the International Committee on Radiation Protection's (ICRP) standards for radiation risks: "The mutagenic effects of radiation determined by Chernobyl studies are actually 1000 times higher than the ICRP risk model predicts."[22] Busby and his colleagues found that "the ICRP risk model based on external exposure cannot be used to estimate internal exposure risks." In other words, extrapolations from Hiroshima and Nagasaki are not a valid basis for understanding cancer rates after the Chernobyl reactor accident and radiation releases.

The ECRR report followed on the BEIR VII report, in July 2005, which concluded that there is "no safe level of exposure" to radiation, and finally admitted, as Leuren Moret says, that "very low levels are more harmful per unit of radiation than higher levels of exposure, also known as the 'supra-linear' effect." As I discuss in chapter 3, independent radiation researchers, not associated with the nuclear industry or regulatory system, have been asserting this since the 1970s.

The controversy over low-level radiation first encountered in the CLBI has come full circle, returning to haunt us as DU contamination from Saskatchewan uranium exports makes its way into human lungs out of sight and mind. Though there has been no mushroom cloud, the magnitude of the DU weaponry used since the first Gulf War dwarfs the radiation from the atomic bombs dropped on Japan. Leuren Moret states that a Japanese physicist has calculated that "the atomicity equivalent of at least 400,000 Nagasaki bombs has been released into the global atmosphere since 1991, from the use of depleted uranium munitions." If you are like me, you'll be hoping this is wrong. The casualties of this magnitude of contamination would not die in a single firestorm but from the insidious lingering effects of low-level radiation.

Since 1991 there has been a regional Middle Eastern increase in infant mortality and melanoma, which may be related to the enormous releases of

DU radioactivity. Leuren Moret, a radiation and public health researcher who has done work for the U.N. on the illegality of DU weapons and presently works as Environmental Commissioner for the city of Berkeley, California, says even "alarming global increases in diabetes" may be related to "a rapid response to internal depleted uranium exposure." Definitive epidemiological research on this has yet to be done and is being blocked by military-industrial officials in the U.S., who are apparently trying to protect the nuclear industry from its own lies.

The withholding of the Aldermaston radiation measures by Halliburton wasn't the first attempted cover-up of the environmental impact of DU weapons. The WHO was involved in a cover-up of its own, refusing to release a report warning of the dangers of low-level radiation from DU weaponry, which was finally leaked and received coverage on September 14, 2004, on the Aljazeera Television program "Washington's Secret Nuclear War."

Chris Busby has published his own assessment of the implications of this uranium contamination on Britain's civilians. He says:

> On impact, uranium penetrators burn fiercely to give an aerosol of sub micron diameter oxide particles which are largely insoluble and remain in the environment for many years.... The question of the dispersion of uranium aerosols from the battlefield is of significant legal interest, since if a radioactive weapon resulted in the general contamination of the public in the country of deployment or elsewhere, the weapon would be classified as one of indiscriminate effect.[23]

Based on the mean increase found in uranium in the air he calculates that "each person in the area inhaled some 23 million uranium particles."[24]

Is this a lot? Inconsequential? If, as the BEIR VII now accepts, there is no safe level of radiation exposure, then DU weapons are not only putting people in Iraq at risk. These weapons infest the lungs of soldier and civilian, friend and enemy, near and far. Those targeting the demonized "other" with DU weapons are endangering themselves. The whole system of warfare breaks down in this ecological interconnectedness. The only thing standing between this realization and governments being held accountable — including uranium-supplying governments like Saskatchewan — is secrecy, deceit and persisting public amnesia.

Previous studies collected evidence of DU contamination in Kosovo and Bosnia. However, this newer evidence from Britain suggests that we are really at war with ourselves. Furthermore, evidence from Iraq, in cases such as the attack on Fallujah in 2004, which Leuren Moret chronicles, suggest "that the U.S. has unethically and illegally used depleted uranium munitions on cities and other civilian populations." As Busby reminds us, this is a violation of a number of agreements in international law, including those banning weapons having indiscriminate effects. Furthermore, since DU weaponry meets most of the criterion of being a WMD, we could say the U.S. did finally find WMDs in Iraq — its own!

The documentary *Beyond Treason* chronicles the impact of DU on both American troops and Iraqi civilians after 1991. Another, *Blowing in the Wind*, chronicles the impact of DU after the U.S. bombing range was moved from Puerto Rico to Australia. Australia is presently being targeted for more than DU weapons testing. Already producing 20% of the world's uranium, second only to Saskatchewan, it is estimated that $36 billion worth of uranium deposits are being targeted for extraction over the next decade. Some uranium mining has continued there since the 1980s, but new *in situ* leach mining has brought production costs down and high uranium prices have made the lower-grade uranium "economic," i.e., profitable

Australians are now following in the same path that Saskatchewan has taken, with all its deadly secrets. A thousand-mile railway stretching from the uranium deposits in the interior to the north coast — likely to sell uranium to China — has already been constructed by none other than Halliburton, the U.S. corporation previously headed by U.S. Vice President Cheney. As the quickly integrating uranium-nuclear industry tightens it global grip, the world seems to be getting smaller even as the nuclear threat enlarges.

Challenges to Saskatchewan's Conscience

Even after years of intense engagement on the uranium-nuclear controversy, I still find it saddening to talk of this deplorable heritage of complicity. I was raised to believe in the "common wealth" in the Cooperative Commonwealth Federation (CCF). As a child I thought this meant we were all — all of humanity — in a big community. So the fact that parochial self-interest and denial can blind us from what the uranium industry is doing to the earth and will do to future generations is a little hard to swallow.

As I was coming to the ending of this book I had many fond and vivid memories and feelings about my late father Bill — my most intimate activist/colleague on this complex "issue." I was even a bit excited at the thought of some closure and lifting of the burden that this project has often carried, especially since attending the World Uranium Hearings in 1992 and learning of the global scope of nuclear devastation. I know how hard it was for my father to break from the NDP over the matter of uranium mining. He was after all, like my late mother Bea, a lifelong supporter of the CCF-NDP. After the CCF was elected in 1944, he moved our family from Alberta to Saskatchewan so he could work in the greater public interest as an agricultural public servant and adult educator. Prior to the Medicare crisis, he acted as provincial secretary of the NDP. The idealism that took him and my mother on their global journeys with the U.N. was spawned within the progressive community around the CCF-NDP. But social democracy based on contamination and proliferation was "not on"; and he made the right choice, giving his last years to help nurture a new vision of sustainability and justice, which the next generation of activists must continue developing and realizing.

I, too, have had to face up to the destructive nuclear path that Saskatchewan NDP governments have followed for decades and to finally and loudly say "No!" However, until I undertook to put the pieces together to tell this long story, my deeper voice was quelled. Though it has been difficult to uncover and fully digest the very dark side of nuclear politics in Saskatchewan since the 1970s, that is what I have set out to do. As I come to the end, with spring energy everywhere, I yearn to "get back to the garden," and I mean this in both a literal and metaphorical sense.

There seems to be a strand that helps explain how a well-intentioned political movement like the CCF-NDP could end up embracing and rationalizing such destruction. At the largest Regina march against the U.S. invasion of Iraq in March 2003, I noticed several high-profile NDPers walking along side the rest of us. One person, who was in both Romanow's and Calvert's cabinets, came up to speak to express opposition to the war. I used to discuss "uranium" with this minister prior to the decision to enter politics. Once nominated and elected, a "pragmatic" choice was made to accept the government's policy supporting uranium mining and apparently to never look back. Now, here was this NDP cabinet minister protesting the war on Iraq — a war that has brought great destruction to the land and people with the aid of massive DU bombs that certainly have the stamp of Saskatchewan on them. I could barely contain the dissonance I felt as the NDP MLA went on about the injustices of the war — treating it as an external political event, far way from us, which invited a righteous response. But I constrained myself.

What was happening? Was I observing completely bifurcated consciousness? Somewhat like Tommy Douglas speaking for nuclear disarmament while knowing that Saskatchewan uranium was fuelling the nuclear arms race? Perhaps a little like supporting uranium exports, so people elsewhere can create spent fuel that will be toxic for 800 generations, while saying "no nuclear here… not in my back yard."

Somehow righteousness nurtures this compartmentalization. Definitely it is an "out of sight, out of mind" approach to politics and morality, which we are quickly learning — with the aid of climate change — is a roadmap to collective disaster. It is time to wake up. Together people do make historical change. Perhaps we need to stop dwelling so much on injustices abroad, as Canadians have a big historical job to do for future generations right here at home.

There are profound ethical, ecological and political questions with which we must grapple. The role of the state in the nuclear industry must be diagnosed and named. Federal and Saskatchewan-based Crown corporations have played a central role in nuclear proliferation. The AECL and its Candu reactor has been kept afloat by ongoing federal subsidies, and Cameco, the main Saskatchewan supplier of DU to the U.S., wouldn't exist if it were not for the huge government subsidies that went to develop the Crown corporations, Eldorado Nuclear and the SMDC, from which it was formed.

These interrelationships fundamentally discredit the nuclear regulatory system. All regulatory bodies, whether Canadian (AECB, CNSC) or American (NRC) or even international (IAEA), are unapologetically pro-nuclear. Personnel often go back and forth between regulatory bodies, the Crowns and private nuclear corporations, all working from the same myopic vision. The Canadian government shields the nuclear reactor industry from total liability, creating the illusion that it is competitive through the *Nuclear Liability Act*, while the IAEA discourages its sister organizations, the World Health Organization and United Nations Environmental Programme, from monitoring the health effects of DU contamination.

The nuclear/uranium industry has operated in a blatantly anti-democratic and secretive manner, and the struggle for sustainability will require a renewed social movement for democratization. For the most part the Canadian public still doesn't know much about this hidden, manipulated history. This of course goes beyond the nuclear industry and its threats to public health and ecology. Between 1941 and 1945, 2,500 young soldiers were exposed to chemical weapons at the CFB Suffield (Alberta) weapons base, and it took fifty years, long after the public could intervene, for this information to be declassified. The *War Measures Act* protected the nuclear industry from public scrutiny during its earliest years, when even the Ban-the-Bomb nuclear disarmament movement knew nothing of the secret uranium military exports for the U.S. and U.K. As so movingly shown in the documentary *Village of Widows*, Dene people, whose close relatives worked in the Port Radium mine and died in excessive numbers from lung cancer, are only now beginning to understand that they, along with the Japanese, were victimized by the Manhattan Project.

Has anything fundamentally changed? Most people in Saskatchewan probably don't yet know about the uranium industry's connection to the first nuclear arms race (1953–66), let alone the second (1980s–present). And the denial we hear from politicians and corporate spokespeople about any connections to the DU weapons used in Iraq and elsewhere are shameless repeats of past lies claiming that Saskatchewan uranium didn't go into nuclear arsenals. When NDP Industry Minister Eric Cline was speaking to the World Nuclear Association in London in 2004, it wouldn't have crossed his mind that had he been there a year before he might have breathed in uranium aerosols from DU weapons made in part from the Saskatchewan uranium he was bragging about at the meeting — weapons dropped on Iraq and brought to England care of Mother Earth's global wind streams. But he'd have no way of knowing, even a year after the events, with Halliburton attempting to keep this information from the public.

These deadly secrets show that the legalistic protections against radioactive pollution and proliferation are not effective. There is steadily growing documentation by the IAEA of uranium "gone missing" — in 1965, 1968, 1973, 1976, 1978, etc. — long before the threat of "terrorists" getting hold of weapons-grade material became such a political football. Though Canada has officially banned exports of nuclear technology or uranium for military purposes since the NPT in 1970, we know that nuclear and uranium industry activities since then have contributed directly to nuclear proliferation. This is clearly true for Candu technology exports, but it is also true for ongoing uranium exports from Saskatchewan to the U.S., France and other countries.

It is time we faced the truth. We are at a juncture: the threat of global warming and the challenge of the Kyoto Accord are fundamentally related to the possibility of further wars due to fossil fuel-dependent militaristic industrialization — underscoring the necessity for social and technological transformation and economic conversion to an ecologically sustainable society. This time we have to do it right, and that means that new energy technologies must pass the ultimate test of sustainability. Uranium-nuclear flatly fails.

Meanwhile, the nuclear industry is trying, once again, to manipulate a comeback. It mostly failed to exploit the OPEC oil-price energy crisis in the 1970s but is now trying to play an "environmental card," saying, as the CNA claims in its new rash of TV ads, it is the "clean, affordable and reliable" alternative to oil and global warming. The absurdity of this nuclear propaganda, that somehow nuclear is a magic bullet to combat global warming, is shown when you realize that if a Candu were built in Saskatchewan it wouldn't even be on SaskPower's grid, but would likely end up providing energy to bring the heavy oil reserves in Alberta onto the world oil market. The non-renewable oil and nuclear industries make good partners, and radioactive pollution is not a positive alternative for reducing greenhouse gases.

Meanwhile, the nuclear/uranium industry is changing. Cameco started as a Crown corporation (the SMDC) that was to bring the benefits of uranium mining to the people. This policy, originating from the Blakeney NDP, failed miserably. Now, as a multinational corporation, Cameco owns uranium mines in northern Saskatchewan and a uranium refinery, conversion plant and nuclear power plants in Ontario, in addition to ventures abroad, including uranium exploration in Australia since 1992. This puts it, along with the French company Cogema (Areva) — the other big mining corporation operating in the north — among the most integrated uranium/nuclear corporations on the planet. Of course to ensure there is growing demand for uranium, it will support both a uranium refinery and a nuclear reactor in Saskatchewan. And of course it would like a nuclear waste storage industry somewhere, such as in Saskatchewan, to off-load nuclear wastes, accumulating from Toronto to Paris, and to profit even more from the complete nuclear fuel system.

Unfortunately, some of our politicians, including many NDP government politicians, are not yet on a learning curve about these matters. So we face more of the same old Orwellian double-speak that the industry and its government backers have used from the beginning: we simply have to learn to accept that uranium mine tailings, nuclear reactor wastes, DU contamination and escalating human cancers are "clean"; that capital-intensive centralized structures requiring huge public subsidies and debt for construction and decommissioning, and becoming less reliable with aging, are "affordable"; that renewable energy alternatives have no big place in the market; and that high-cost, inefficient electrical generation, nuclear proliferation and the spreading threat of nuclear war will make our future "reliable."

Realistically, time does not always seem to be on the side of sustainability, though it's always on the side of evolution. Luckily, we are already imagining a journey of transformation towards sustainability — something that is cultural and spiritual, as well as technological and political. And it must also be economic. It is notable that appealing to the economic ideology ("corporate profit and capital-intensive jobs at any cost") that greatly got us into this military-industrial-ecological mess in the first place is the last line of defence for the expansion of the nuclear system in Saskatchewan. This ideology helps maintain the amnesia and hide the deadly secret about what the uranium/nuclear industry is actually doing to the planet and its creatures. It's time to act out of self-defence as well as compassion; to act mindfully as well as militantly; and quickly to do our part to redirect the Saskatchewan region from being the front-end of the deadly global nuclear system.

Notes

1. Fraser Needham, *The Commonwealth*, September–October, 2006, Vol. 66, No. 4, p 1–2.
2. See Ernie Regehr's *Arms Canada*, Toronto, Lorimer, 1987, and more recent writings from Project Ploughshares. Up-to-date information on Canada's role in missile defence is in Steve Staples, *Missile Defence: First Round*, Toronto, Lorimer, 2006.
3. Yussuf Kly, *The Regina Seminar on the Elimination of Weapons of Mass Destruction*, Altanta, Clarity Press, 2006.
4. Jim Harding and Dave Gullickson, *Provincial Policy Chronology of Uranium Mining in Saskatchewan*, Report No. 3, In The Public Interest Series, Prairie Justice Research, University of Regina, Nov. 1997, pp. 1–20.
5. Carole Giangrande, *The Nuclear North*, Toronto, Anansi Press, 1983, p. 98.

6. See Ronald Babin, *The Nuclear Power Game*, Montreal, Black Rose, 1985, p. 44.
7. Quoted in Rod Nickel, "Sask. uranium used for energy, not bombs," Saskatoon *Star Phoenix*, March 25, 2006, p. E1.
8. See S. Weissman and H. Krosner, *The Islamic Bomb*, New York Times Books, 1991, p. 91.
9. For a discussion of the Uranium Club, which included Canada, see Jim Harding, *Social Policy and Social Justice*, Wilfrid Laurier University Press, 1995, pp. 348–53.
10. David Fleming, "Why nuclear power cannot be a major energy source," April, 2006, p. 6, available at <http://www.feasta.org/documents/energy/nuclear_power.htm> accessed June 2007. The figure will be higher now.
11. Leuren Moret, "Depleted uranium: The Trojan Horse of nuclear war," *World Affairs: The Journal of International Issues*, July 2004, available at <http://www.globalresearch.ca/index.php?context=va&aid=709> accessed June 2007.
12. Calculated from Cameco.com.
13. Cameco Quarterly Report, July 28, 2006, p. 21. Also personal communication by phone with Cameco staff.
14. Mineral Statistics Section, Mines Branch, Saskatchewan Industry and Resources, November 9, 2006. I added 30.0 million pounds, an average for recent years, for both 1998 and 1999, which had missing data due to only two corporations being active.
15. The Coalition to Oppose the Arms Trade <http://coat.nfc.ca> is a good source of information. Also, see Leuren Moret, "Depleted Uranium: The Trojan Horse of nuclear war."
16. In a letter to the ICUC dated October 30, 1990, Saskatchewan Environment and Public Safety admitted that in 1989 some yellowcake from SMDC went to Gore Oklahoma where Sequoyah Fuels is based.
17. ICUC, press release, January 21, 1993. Also see Phillip Penna, "Cameco and its weapons connections," *Briarpatch*, May, 1993, p. 19.
18. In "Current Issues: Waste Management of DU," posted at *WISE-uranium.org/DU*, updated November 20, 2006, there is information about Manufacturing Science Corporation of Oak Ridge applying to export 10.21 metric tonnes of U308 to Canada as "test material to make DU oxide in the Cameco Corporation facilities in Port Hope, Ont." This licence was issued March 17, 2004.
19. Gordon Edwards, "DU Munitions," email, Sept. 13, 2006.
20. For more details see S.N. Al-Azzawi, "Depleted Uranium Contamination in Iraq: An Overview," presented to 3rd ICBUW [International Coalition to Ban Uranium Weapons] International Conference, Hiroshima, Aug. 3–6, 2006.
21. These findings were reported in Mark Gould and Jon Ungoed-Thomas, "UK radiation jump blamed on Iraq shells," *Sunday Times*, February 19, 2006.
22. Leuren Moret, "From battlefields in the Middle East: Depleted uranium measured in British atmosphere," March 2, 2006, available at <http://www.globalresearch.ca/index.php?context=va&aid=2058> accessed June 2007.
23. Chris Busby and Saoirse Morgan, "Evidence from the measurements of the Aldermaston Weapons Establishment, Aldermaston, Birkshire, U.K." Occasional Paper 2006/1, Green Audit, Jan. 2006.
24. The mean increase in uranium was calculated to be 500nBq/m3, an estimate of Bequerels of radiation per cubic metre, and the size of uranium particles inhaled was calculated as 0.25 microns.

ACRONYMS

AEC	Atomic Energy Commission
AECB	Atomic Energy Control Board
AECL	Atomic Energy of Canada Limited
AIM	American Indian Movement
AMNSIS	Association of Métis and Non-status Indians of Saskatchewan
AOT	Aerojet Ordnance Tennessee
AWE	Atomic Weapons Establishment
BCMA	B.C. Medical Association
BEIR	Biological Effects of Ionizing Radiation
BOG	Board of Governors
CANDU	Canadian Deuterium Uranium
CCF	Cooperative Commonwealth Federation
CCNR	Canadian Coalition for Nuclear Responsibility
CEA	Commissariat de l'Energie Atomique
CEGB	Central Electricity Generating Board
CEO	chief executive officer
CFB	Canadian Forces Base
CLBI	Cluff Lake Board of Inquiry
CLJV	Cigar Lake Joint Venture
CNA	Canadian Nuclear Association
CNSC	Canadian Nuclear Safety Commission
CRISLA	Chemical Reaction by Isotope Selective Laser Activation
CUCND	Combined Universities Campaign for Nuclear Disarmament
DIAND	Department of Indian Affairs and Northern Development
DNS	Department of Northern Saskatchewan
DOE	Department of Energy
DSM	demand-side management
DU	depleted uranium
ECRR	European Committee on Radiation Risk Report
EDF	Electricite de France
EIS	Environmental Impact Statement
EMR	Energy, Mines and Resources
EPA	Environmental Protection Agency
ERP	Environmental Review Panel
FEARO	Federal Environmental Assessment Review Office
FIRA	Foreign Investments Review Agency
FSI	Federation of Saskatchewan Indians
FTA	Free Trade Agreement
GDP	gross domestic product
GHG	Greenhouse gases
GNP	gross national product
GNWT	Government of the N.W.T.
IAEA	International Atomic Energy Agency
ICRP	International Committee on Radiation Protection
ICUC	Inter-Church Uranium Committee
ISM	Information Systems Management
IUC	International Uranium Congress
JFPP	Joint Federal-Provincial Panel
KIA	Keewatin Inuit Association
KLBI	Key Lake Board of Inquiry

KLMC	Key Lake Mining Corporation
KRC	Keewatin Regional Council
KRICC	Keewatin Regional Intervention Coordinating Committee
MLA	Member of Legislative Assembly
MOU	Memorandum of Understanding
MP	Member of Parliament
NAFTA	North American Free Trade Agreement
NATO	North Atlantic Treaty Organisation
NDB	Northern Development Board
NDP	New Democratic Party
NEA	Nuclear Energy Agency
NEB	National Energy Board
NFB	National Film Board
NGO	non-governmental organization
NIRS	Nuclear Information and Resource Service
NMC	Northern Municipal Council
NPT	Non-Proliferation Treaty
NRC	Nuclear Regulatory Commission
NWMO	Nuclear Waste Management Organization
OECD	Organisation for Economic Co-operation and Development
OPEC	Organization of Petroleum Exporting Countries
OPIRG	Ontario Public Interest Research Group
PIU	Performance and Innovation Unit
PJR	Prairie Justice Research
RGNNS	Regina Group for a Non-Nuclear Society
SARM	Saskatchewan Association of Rural Municipalities
SCIC	Saskatchewan Council for International Co-operation
SDU	Students for a Democratic University
SECDI	Saskatchewan Energy Conservation and Development Institute
SED	Space Engineering Division
SES	Saskatchewan Environmental Society
SF	Sequoyah Fuels
SMA	Saskatchewan Mining Association
SMDC	Saskatchewan Mining and Development Corporation
SNPP	Student Neestow Partnership Project
SRC	Saskatchewan Research Council
SSHRC	Social Sciences and Humanities Research Council
SUMA	Saskatchewan Urban Municipalities Association
SUPA	Student Union for Peace Action
SUV	sports utility vehicle
SWU	separate work units
TRW	Thompson-Ramo-Wooldridge
UHF	uranium hexafluoride
UNDP	United Nations Development Programme
UNEP	United Nations Environment Programme
USAEC	U.S. Atomic Energy Commission
USEC	U.S. Energy Corporation
WCED	World Commission on Environment and Development
WHO	World Health Organization
WLJV	Waterbury Lake Joint Venture
WMD	weapons of mass destruction
WPDA	Western Project Development Associates
WTI	World Tribunal on Iraq

CHRONOLOGY OF SIGNIFICANT EVENTS, 1942–2007

Note: This chronology will help the reader of *Canada's Deadly Secret* create historical context — from the local situation to the global — and provide more familiarity with the complex and potentially confusing litany of organizations involved in nuclear politics in Saskatchewan, Canada.

The Cold War and Nuclear Arms Race: SK as Supplier

1942 Eldorado Nuclear's Port Radium mine in N.W.T. reopens to provide uranium for Manhattan A-bomb Project

1943 Canada becomes full-fledged member in Manhattan Project; Depleted uranium (DU) weapons first envisaged by Manhattan Project

1944 Eldorado Nuclear becomes federal Crown corporation; Co-operative Commonwealth Federation (CCF) led by Tommy Douglas first elected in SK

1945 Chalk River, Ontario, reactor starts producing weapons-grade plutonium for U.S.; U.S. drops A-bombs on Hiroshima and Nagasaki

1946 Canadian Atomic Energy Control Board (AECB) Act passed; Eldorado Nuclear stakes uranium claims at Beaverlodge, SK

1951 U.S. Atomic Energy Commission (USAEC) guarantees market and profitable price for all Eldorado Nuclear uranium

1952 U.S. tests first H-bomb

1953 Beaverlodge, SK, uranium mine opens; U.S. President Eisenhower makes "atoms for peace" speech

1957 major nuclear reactor accident at Windscale, England

1958 Ban-the-Bomb protests of nuclear arms race begin in SK

1961 Tommy Douglas elected leader of newly formed New Democratic Party (NDP); Woodrow Lloyd elected SK NDP leader-Premier elect

1962 Eldorado Nuclear's first public admission of role in Manhattan Project; Medicare established in SK

1963 Pearson federal Liberal government does turn-about and supports nuclear-armed Bomarc missiles on Canadian soil; NDP leader and Member of Parliament Tommy Douglas supports peace movement's protests; Limited Test Ban Treaty (there is still no comprehensive treaty)

1964 Ross Thatcher Liberals defeat Woodrow Lloyd NDP in SK provincial election

1965 SK Thatcher Liberal government fails to get heavy water plant at Estevan

1967 French uranium corporation Amok begins aerial surveys of northern SK

1968 Gulf Minerals discovers uranium deposit at Rabbit Lake, SK

1969 German corporation Uranerz guarantees market for all Rabbit Lake uranium

1970 Non-Proliferation Treaty (NPT) passed

The Energy Crisis: SK Sticks with Uranium

1971 Allan Blakeney NDP defeats Thatcher Liberals in SK provincial election; high-grade uranium deposits discovered at Cluff Lake; SK government fails to get uranium enrichment plant for Estevan

1972 Atomic Energy of Canada Limited (AECL) sells Candu reactor to India; then to Pakistan; uranium cartel, which includes Canada, inflates uranium prices until 1974

1973 SK NDP government fails to get heavy water plant at Estevan; AECL attempts Candu sale to Argentina's military government; Organization of Petroleum Exporting Countries (OPEC) raises price of crude oil and "energy crisis" begins; oil-import-dependent Japan expands nuclear power

1974 Blakeney NDP government creates public uranium corporation, Sask. Mining and Development Corporation (SMDC); AECL almost sells Candu to Iraq

1975 high-grade uranium deposits found at Key Lake, SK

1976 uranium issue erupts at SK provincial NDP convention — uranium inquiry, not moratorium, accepted as compromise; Eldorado Nuclear proposes uranium refinery for SK

1977 Regina Group for Non-Nuclear Society (RGNNS) forms; construction begins at Amok and SMDC-owned Cluff Lake uranium mine; Cluff Lake Board of Inquiry (CLBI) begins hearings; all SK First Nations and Métis organizations call for moratorium on uranium mining; SK major uranium customer, France, continues with nuclear weapons tests (until 1996)

1978 SK Heritage Fund created as contingency fund for unforeseen environmental problems with uranium mining (squandered by 1991); Ontario's "Porter" Commission on Electric Power Planning calls for moratorium on nuclear power until waste problem resolved; CLBI final report endorses uranium expansion in SK

1979 SK NDP government creates Uranium Secretariat, which launches several disinformation projects; RGNNS publishes Bill Harding's *Correspondence with the Premier*; leaked SK Department of Environment memos confirm illegal lakes drainage at Key Lake; major nuclear accident at Three Mile Island, Pennsylvania

1980 SK Key Lake Board of Inquiry (KLBI) begins hearings; Inter-Church Uranium Committee (ICUC) forms in Saskatoon; Federal Environmental Assessment Review Organization (FEARO) Panel hears and rejects proposal for uranium refinery at Warman, SK; RGNNS publishes *Why People Say No!* (to uranium refinery); British Columbia declares seven-year moratorium on uranium mining; referendum on nuclear phase-out held in Sweden

1981 Saskatoon Atoms for War and Peace conference reveals SK's continuing connections with nuclear weapons; Key Lake Mining Corporation (KLMC) charged with illegal drainage; KLBI issues final report; AECL's largest Candu sale ever made to Communist Romania; Ronald Reagan elected U.S. President — escalates second nuclear arms race, which increases demand for uranium security of supply

Nuclear Globalization: SK Becomes the Front-End

1982 Grant Devine Tories defeat Blakeney NDP in SK provincial election; Sask Power's Office of Energy Conservation shut down

1983 highest-grade ever uranium discovered at Cigar Lake, SK; in opposition SK NDP reverses policy and endorses uranium mining phase-out; Prairie Justice Research at University of Regina receives Social Science funding to undertake Uranium Inquiries Project (publications under Series In The Public Interest continue until 1997)

1984 major uranium spill at Key Lake SK uranium mine; Brian Mulroney Tories elected as federal government

1985 U.S. Cruise missile testing begins in western Canada

1986 catastrophic nuclear accident at Chernobyl, Ukraine — radioactive contamination of many countries

1987 Inter-Church Uranium Committee (ICUC) campaigns to make Saskatoon a Nuclear Weapons Free Zone; Canadian Nuclear Association (CNA) targets SK in Canada-wide nuclear ad campaign; SK networking with European Greens begins; United Nations World Commission on Environment and Development (Brundtland) Report on sustainable development endorses renewable energy

1988 SMDC and Eldorado Nuclear merged and privatized as Cameco; Greenpeace campaign against Cigar Lake; International Uranium Congress (IUC) organizing committee formed and event held in Saskatoon; U.S. Supreme Court rules against domestic uranium protectionism; Free Trade Agreement (FTA) signed between Canada and U.S.

1989 major uranium spill at Wollaston Lake, SK; widespread calls for new SK uranium inquiry; huge shortfall on projected SK uranium revenues; Inuit at Baker Lake, N.W.T, vote 90% against uranium mine in vicinity; Western Project Development Association (WPDA) proposes Candu reactor for SK; Energy Probe study shows growing costs of aging Candus; CNA releases "Nuclear Facts" promotional pamphlets; Economic Council of Canada estimates $12 billion in subsidies to AECL

1990 AECL head speaks at University of Regina, Engineering series; Saskatchewan Education promotes *Uranium* in its Resources Series; SK Department of Economic Diversification and Trade begins talks over Candu-3 with TRW (Thompson-Ramo-Wooldridge); National Film Board releases its film *Uranium* in face of stiff opposition within federal bureaucracy; new Ontario Rae NDP government begins investigation of Ontario Hydro; first Candu sale in a decade — to South Korea; OECD's Nuclear Energy Agency (NEA) reports electricity from nuclear only 26% of that forecasted by it in 1972

1991 SK Devine government signs Memorandum of Understanding (MOU) with AECL re Candu-3; Roy Romanow NDP defeats Devine government in SK provincial election; Sask Power's Energy Options Panel issues final report — supports conservation; NDP convention rejects Candu reactor; National Energy Board (NEB) forecast contradicts AECL on projected growth in SK electrical demand; SK Energy and Mines releases "Uranium in SK" promotional educational series; SK's Gass Report documents writing down of value of Cameco's public share offering; CNA holds annual meeting in Saskatoon; AECL and Cameco propose SK as nuclear waste site; FEARO Panel on AECL's nuclear waste proposals begins hearings; Darlington Candu plant cost overruns hit $11 billion; Chernobyl clean-up costs at $18 billion; *The Economist* reveals huge French nuclear reactor debt load; DU weapons first used on occasion of Gulf War

1992 SK NDP government Deputy Leader visits TRW in Washington; Joint Federal Provincial Panel (JFPP) on five new SK uranium mines begins hearings; SK NDP convention supports uranium mining expansion only if approved by environmental review; Romanow launches SK Energy Conservation and Development Institute (shut down in 1996); Romanow NDP signs new MOU with AECL and AECL moves Candu-3 offices to Saskatoon; Saskatchewan Mining Association (SMA) releases teacher's guide *Uranium in SK*; Earth Summit held in Brazil; 1600 scientists issue global warning about devastating ecological impact of energy-intensive industrial society; France and China sign NPT; World Uranium Hearings held in Salzburg, Austria

1993 JFPP's first report rejects one new SK uranium mine, calls for postponement of another and warns no guarantee SK uranium used only for non-military purposes; ICUC releases evidence that DU from SK is used in U.S. weapons; newly elected Chrétien federal Liberal government cancels Tories nuclear submarine deal; U.S. reports finds that nearly $500 billion of public finding has gone into nuclear industry

1994 Romanow NDP government ignores JFPP and approves new uranium mines; SK NDP government makes initial sale of Cameco stocks; Cameco launches aggressive northern and Saskatoon school and community promotions; North American Free Trade Agreement (NAFTA) signed between Canada, Mexico and U.S. (uranium agreements of FTA continue)

1995 SK secular non-nuclear movement hibernates

1997 Kyoto Accord on climate change initiated; promotion of nuclear as "clean" energy intensifies

1998 SK Green Party begins to field candidates — advocates non-nuclear options; FEARO Panel on AECL nuclear waste proposal reports without any practical solutions

1999 Romanow NDP re-elected as minority government forms coalition with Liberal MLAs to hold provincial power; ICUC sues AECB over Cogema licence at McLean Lake, but is later refused standing to appeal case at Supreme Court

2000 Canadian Nuclear Safety Commission (CNSC) replaces AECB as federal regulatory body; NEA reports that 14% of Canadian electricity comes from nuclear (among lowest in OECD)

New Directions: The Challenges of Sustainability

2001 Lorne Calvert elected SK NDP leader-premier elect; Cameco takes control of privatized Bruce Power Candus in Ontario; first wind farm in SK; French government renames uranium/nuclear conglomerate Areva; George W. Bush elected U.S. president; 9/11; DU weapons used in Afghanistan

2002 SK NDP government sells remaining Cameco shares; SK NDP government lowers fossil fuel royalties and along with Alberta stalls on Kyoto Accord; provincial NDP convention supports Kyoto Accord; industry-based Nuclear Waste Management Organization (NWMO) established by federal Liberal government

2003 SK Calvert NDP barely re-elected — 30 of 58 seats; paper at Canadian International Petroleum Conference proposes Candu for Alberta tar sands; DU weapons used in the invasion and occupation of Iraq; uranium aerosol count rises as far away as England

2004 Cigar Lake mine licensed; SK Minister of Industry speaks at World Nuclear Association; University of Regina Justice Studies Roundtable on eliminating WMD; United Nations Environmental Programme (UNEP) finds energy subsidies still biased towards nuclear and fossil fuels; DU weapons likely used in U.S. attack on city of Fallujah, Iraq

2005 SK Centennial Action Plan for Economy supports uranium industry; Calvert NDP government and party supports uranium refinery; NWMO holds public consultations in SK about long-term nuclear waste disposal; George Bush regime creates $13 billion in new subsidies for nuclear industry; Kyoto Accord comes into effect; New Economics Foundation and *New Scientist* report cost of nuclear power underestimated by factor of three; World Tribunal on Iraq (WTI), concluding in Turkey, hears evidence of cancer increases from DU contamination in Iraq

2006 University of Regina co-sponsors "Exploring Saskatchewan's Nuclear Future" conference; massive flooding in Cigar Lake mine tunnels — start-up postponed; Suzuki Foundation finds SK greenhouse gases grown most and highest (per capita) in Canada; Calvert NDP appoints Legislative Secretary for Renewable Energy — two reports later released; Calvert visits U.S. Vice President Cheney to promote non-renewables uranium and oil; Calvert promotes uranium refinery while visiting France; Energy Probe estimates subsidies to Canadian nuclear industry total $75 billion; AECL announces deal with Alberta Energy Corporation to promote Candu in oilfields; Ontario McGinty Liberal government announces plans for two more reactors at Darlington; Harper Conservative minority government elected: rejects Kyoto targets and considers selling AECL; BEIR VII (Biological Effects of Ionizing Radiation) concludes no safe level of radiation; uranium price almost doubles, spearheading uranium exploration worldwide; China negotiates uranium supply with Australia; Helen Caldicott publishes *Nuclear Power Is Not the Answer* (for global warming)

2007 SK Environmental Society releases non-nuclear *Sustainable Energy Strategy for SK;* Helen Caldicott's speaking tour in SK; Calvert NDP brings Al Gore to speak in Regina; Calvert NDP releases *Green Strategy* and *Sustainable Energy Strategy* — both propose "energy mix" that includes uranium/nuclear; SK provincial election pending — polls put Calvert NDP far behind Sask Party; Alberta Energy Corporation announces plan to file Candu site application; new SK Non-Nuclear Network starts lobby of all MLAs; ecumenical activist coalition KAIROS focuses on non-nuclear options at Prairie Conference on Sustainable and Just Energy.